INSECT
WORLDS

BY LORUS J. MILNE AND MARGERY MILNE

FOR THE GENERAL READER

Insect Worlds
The Audubon Society Field Guide to North American Insects and Spiders
Ecology Out of Joint
The Secret Life of Animals: Pioneering Discoveries in Animal Behavior (with Franklin Russell)
The Animal in Man
The Arena of Life: The Dynamics of Ecology
Invertebrates of North America
The Cougar Doesn't Live Here Anymore: Does the World Still Have Room for Wildlife?
The Nature of Life: Earth, Plants, Animals, Man, and Their Effect on Each Other
North American Birds
Patterns of Survival
Living Plants of the World

The Ages of Life: A New Look at the Effects of Time on Mankind and Other Living Things
Water and Life
The Valley: Meadow, Grove and Stream
The Senses of Animals and Men
The Mountains (with the Editors of LIFE)
The Balance of Nature
The Lower Animals: Living Invertebrates of the World (with Ralph and Mildred Buchsbaum)
Plant Life
Animal Life
Paths Across the Earth
The World of Night
The Mating Instinct
The Biotic World and Man
A Multitude of Living Things

ESPECIALLY FOR YOUNG PEOPLE

Gadabouts and Stick-at-Homes: Wild Animals and Their Habitats
Because of a Flower
The How and Why of Growing
The Nature of Plants
When the Tide Goes Far Out

The Phoenix Forest
Gift from the Sky
The Crab That Crawled Out of the Past
Because of a Tree
Famous Naturalists
The Nature of Animals

INSECT WORLDS

A GUIDE FOR MAN ON MAKING THE MOST OF THE ENVIRONMENT

Lorus J. Milne

&

Margery Milne

CHARLES SCRIBNER'S SONS/NEW YORK

All photo illustrations in this book are by the authors, except where otherwise indicated in the captions.

Some of our own observations, particularly those on the realm of the water surface and the burying beetle, have been described previously in Scientific American *(1976, 1978).*

Quote from Borne on the Wind: The Extraordinary World of Insects in Flight *by Stephen Dalton,* © *1975, reprinted with permission.*

Copyright © *1980 Lorus J. Milne and Margery Milne*

Library of Congress Cataloging in Publication Data
Milne, Lorus Johnson, 1912–
Insect worlds.
1. Insects—Behavior. I. Milne, Margery Joan Greene, 1914– joint author. II. Title.
QL496.M49   595.7'051   80–16218
ISBN 0–684–16627–5

1 3 5 7 9 11 13 15 17 19 H/C 20 18 16 14 12 10 8 6 4 2

Printed in the United States of America.

FOR
Anthony G. Marshall,
who keeps expanding his interface
with the environment

Contents

INSECT
WORLDS

1

"Ours"
Is an Insect World

THIS WORLD WE CLAIM AS OURS IS AS MUCH A WORLD OF INSECTS. For millions of years these small creatures have outnumbered, in diversity, all other forms of animal life on earth, even though most of them behave in ways that go unnoticed. Compared with insects whose enterprise began about 350 million years ago, in late Devonian times, our ancestors in the human family were relative newcomers who can be traced back only about 3.5 million years. Obviously, then, the survival strategies of insects have been remarkably successful and deserve not only our attention but our admiration.

Although the ways of insects and people are completely different, discoveries about insect life stimulate us to rethink our own behavior. In general, the evolution of insects has resulted in greater diversification and in the minimization of competition among species. By continually finding new opportunities in the environment, they not only expanded hugely in numbers, but were able to share the terrestrial world. Our progress, by contrast, has come about by developing cultural traditions that tend to simplify the environment and eliminate competitors—including insects, if possible. The question yet to be answered is whether our solutions will really ensure the survival of our posterity.

Until insects and their fellow arthropods, the spiders, followed plants into the full sun and existence on land, all life had been aquatic. In fact, the fossil record of Devonian times is so abundantly rich in bony remains of marine and freshwater denizens that we know the period as the Age of Fishes.

During the Carboniferous period (or Coal Age), which followed the Devonian, many insects took to the air and became the first free-flying animals. In the beginning they flew to escape from amphibians—the first vertebrates to survive on land. Eventually, other vertebrates with a palate for insects appeared and had to be avoided. After the amphibians came the reptiles, and following them, about 150 million years ago, the birds—which, along with some of the reptiles, followed insects into the skies. Bats, the only mammals with wings, have pursued flying insects for a mere 60 million years.

What all these precursors had in common, as demonstrated in the fossil record, was an appetite for insects. Thus, we can conclude that insects were largely responsible for making vertebrate life possible on land. But they also changed the plant world, stimulating the evolution of the myriad patterns of plant life we see today.

MARVELS OF MINIATURIZATION

Engineers, particularly those concerned with aeronautics or computer systems, often look enviously at all the actions of which an insect is capable and speak of the little body as a "marvel of miniaturization." But insects have always been comparatively small, even minute. The smallest of them, when fully grown, could walk easily through the eye of a fine sewing needle. The largest today are the goliath beetles of equatorial Africa, which are almost the size of rats, although lighter in weight despite their thick armor covering. A newborn mouse has a larger brain. Yet the limited number of nerve and muscle cells in the insect suffice for the creature to coordinate its activities from its initial wriggling inside the egg until it has contributed to the perpetuation of its species.

Insects eat, work, and sleep. They court and mate. Some tend their eggs and young. They respond to the details of their world, and communicate their needs, their readiness to act, and their awareness of each other. Perhaps more than scientists are yet willing to admit, an insect may be conscious of what it is doing.

Much of the limitless versatility of insects arises from a lucky combination of size and body covering. Small size makes possible rapid growth to mature dimensions, but renders a creature vulnerable to physical damage and loss of vital body moisture into the air. Insects have been protected externally from these hazards since the very beginning by a resistant cuticle secreted by the skin. The cuticle covers not only the outer surface of the body, but lines the digestive tract as well from mouth to anus. Pores perforate the cuticle where exchange of substances in solution is important. Such a cuticle can be shed and replaced quickly at intervals to permit growth. It can also be modified to allow changes as the insect approaches adulthood.

The insect cuticle is mostly protein. Much of the cuticle covering the body surface is stiffened into firm plates by fibrils of a second material with some of the characteristics of fiberglass; this substance is called chitin (pronounced kite'-in), named after chiton, the coat-of-mail armor worn by Greek warriors. Flexibility between the plates is maintained by a distinctive pattern of unstiffened joints between every section of the body except, usually, the wings. All these outer surfaces are coated by a third material, a thin film of fatty substance secreted by the insect that provides a virtually weightless waterproofing. Protected by its special armor, an insect can exploit almost any opportunity. It is able to move about freely, gaining whatever leverage it needs with muscles inside its flexible armor. Depending on how its body covering is specialized by the enlargement of some parts and reduction of others, the insect may creep, run, hop, jump, swim, dive, or fly forward or backward. Some demonstrate a virtuosity of motor control that is truly remarkable.

But no matter how varied in size, shape, or motor behavior, every insect shares the same basic structure. Its body is divided into three major parts: the head, a movable, three-part thorax (or upper body region), and a segmented abdomen. The head is set off with a pair of jointed antennae ("feelers") and paired mouthparts, and the thorax bears up to three pairs of jointed legs and sometimes wings as well. (Spiders differ from insects primarily in having eight legs and only two major divisions of the body.)

Human beings, who already weigh six pounds or more when they first open their eyes to the world, may be tempted to underestimate what a little insect can do with its six-legged body. We tend to regard each beetle or butterfly as though it were some kind of toy with a simple mechanism inside. Smaller than toy-sized it may be, yet each performs routines of surprising complexity and flexibility. As each insect grows, its behavior alters according to an inherited program and also in relation to opportunities in the environment. Male and female may be surprisingly unlike in nutritional requirements and longevity, as well as in their complementary sexual features and actions. However, every one of their actions is a blend of past and present. A brief stimulus of the moment releases behavior that was programmed perhaps millions of generations in the past by the basic struggle for survival.

DEXTERITY, STRENGTH, AND SPEED

Certain insects, even some of those that are most familiar, are remarkably dexterous and move too fast for the eye to follow. One of these is the common housefly. The way a fly alights upside down on a ceiling might still be a secret had not special high-speed motion picture cameras been brought to bear on the subject in 1962. It was then discovered that the housefly normally zooms upward toward the ceiling at 0.56 miles per hour, beating its two wings between 144 and 240 times per second. As the insect comes within its own body length of its target, it reaches forward with its front legs and outward with the middle and hind pairs. The front legs catch the ceiling, and their sticky pads grip firmly. This swings the fly's body around and brings its other four legs into contact with the ceiling. The tricky touchdown stops the fly where it aimed. Such exhibitions of flying were recorded by the English naturalist Stephen Dalton with special cameras and strobe lights, and led him to comment: ". . . the world is a meadow made for insects, not for men. It is a meadow used for takeoff and landing in a nonstop display of flight techniques and aerobatics so stunning and spectacular that, in comparison, even the most advanced manned flight is but the first faltering step of a child."

Left to its own devices, the fly walks about on any surface in the same characteristic gait shown by most other insects. First, the front and rear legs on one side and the middle leg on the other side support the insect. Then its weight shifts to the opposite legs, so that one tripod is always ready to move while the other keeps a good grip on the world.

Dragonflies and damselflies manifest unique and amazing mobility, because their four wings are separately controlled. Each one can be raised and lowered for flapping flight, or moved forward or backward, or tilted on its long axis. The insect shows magnificent control of all these movements, and uses them to hover, to rise or descend, to propel the slender body ahead or to the rear, at almost any angle. Often we see a damselfly demonstrate its graceful maneuvers near the vegetation close to fresh water. Or we delight in watching a dragonfly pursue midges and other prey, or drive another dragonfly from its streamside territory, with dazzling changes in speed and in direction. We rejoice that the world has continued to provide a place for these insects and that their behavior has adjusted so well to changes in their environment, both in the water and above it.

The thrust of the legs on a large insect can be incredibly powerful. We discovered this one day in the Panamanian rain forest, where a huge beetle caught our attention. It was crawling along a short length of fallen tree branch. Getting a good grip on the insect was easy, for its body measured almost two inches across, more than an inch in thickness, and four inches long, including a one-inch projection from the head. We pressed thumb and fingers against opposite sides of the beetle, so as not to get in the way as its six legs continued their normal walking gait. We lifted its body, and the soggy tree branch lifted too, clutched firmly by the sturdy insect. Not until it had clambered to the end of the branch did it drop this former support. Then, with legs no longer engaged in locomotion, the beetle raised them against the fingers holding its body and forcefully push them aside. Its leverage and muscles proved so powerful that only by carrying the beetle by its rhinoceroslike head horn could a hold be maintained. Never have we overcome our astonishment at being unable to restrain an insect because its legs were stronger than our fingers!

In the same rain forest, we learned that large size need not handicap an insect in terms of speed. There, as in similar environments of adjacent South America, some of the world's largest moths are quite at home. Just once have we seen one of these giants alive, but it taught us a lesson. The big insect was clinging to the pale gray bark of a huge tree, perhaps eleven feet above the forest floor. The creature held its body crosswise with its wings spread flat against the trunk, the right pair pointing up, the left pair down. The scalloped margins of the forewings, and of the hind pair that pressed even closer to the tree, blended almost perfectly with the rough surface. It was motionless and appeared to be asleep. Was this the famous moth *Thysania agrippina,* which explorers described as darting about at night like a bat? Its size and form were right, as was its nine- to twelve-inch wingspread.

One of us quietly fitted together the ring and double-extension handle of a butterfly net and suddenly swung the tool to cover the moth. We expected the big insect either to hold its position or to flutter into the conical mesh of the net. With so great a wingspan, *Thysania* might be clumsier than the cecropia moths and related denizens of North American woodlands, where a six-inch wingspread is something of a record. But *Thysania* didn't flutter. The giant moth darted through a two-inch gap between the net ring and the tree, caused by the imperfect fit of the ring against the cylindrical trunk, and propelled itself off through the forest and out of sight at incredible speed. Only much later, in a museum, did we have a chance to examine a specimen of this species, and to see the steel-blue radiance of the wing surface that had been pressed against the bark. Possibly this color is part of another lifesaving trick in the moth's repertoire, a different asset it can call on when its normal camouflage fails to provide concealment.

SENSORY SYSTEMS

All insects keep well informed about external events through sense organs that project from pores in their armor at countless points. Each sense organ is connected to the central nervous system. Tastes and vibrations inform the creature of immediate changes in the environment. Sounds and odors underfoot carry messages from greater distances. Eyes, protected by transparent bulges in the cuticle, vary in complexity: simple eyes can distinguish in light intensity between day and night, sun and shadow, whereas compound eyes are able to recognize colors and patterns and to detect movement with amazing sensitivity.

Hunger, thirst, temperature differences, and sexual readiness are also monitored less conspicuously but no less efficiently. Food holds a high priority in the active life of any insect, at least during the stages of its development between hatching from the egg and attaining sexual maturity. But how much more limited in choosing its foods is a caterpillar or a blowfly than a rabbit, for example? Vincent G. Dethier, a Princeton University physiologist, has compared the fine details of the insect body with those of other animal bodies in an inquiry into dietary habits. A rabbit, he has discovered, has about 100 million special receptors for odors in its nose and throat, with which to analyze each sniff at a leaf. A caterpillar has only forty-eight olfactory receptors, all close to its mouth. A blowfly relies upon 3,120 odor-sensitive bristles on its six legs, plus between 245 and 257 on its mouthparts. Each animal uses its own sensory array to operate amid a sea of olfactory stimuli. Somehow the odors and flavors trigger the correct responses for food and mates and for poisons and other dangers.

Dethier has used primarily the common blowfly for his studies of insect reactions to external stimuli. To date he has found no way to identify the environmental signals that alert the fly as it buzzes about, seemingly at random. Its activity seems to prove that it is hungry, without indicating how the fly detects its nutritional state. Although sensors that monitor its slow-moving blood may report that the concentration of sugar has dropped below a critical level, and pressure sensors may signal that the digestive tract is empty, the fly will keep moving and expending energy from its limited store at the fastest rate. As yet there is no clue to tell whether the energy is still coming from nourishment absorbed during earlier maggothood, or is a dwindling amount left over from the fly's previous meal.

The insect's behavior seems much more understandable after it settles. A fly-sized black spot on any pale surface can serve as a decoy. The spot catches the attention of the airborne fly as a visual pattern. The fly spirals down for a landing, then walks across the surface in short strolls, often grooming itself at each stop. At every step its legs pick up olfactory signals of many kinds, and the tiny brain sorts out the information. Finding a streak of sugar solution, the fly stamps down its soft mouthparts like a damp viscose sponge, ready to mop up whatever is available. Meanwhile, messages continue to come in from its six feet. If one of them chances to step into a droplet with a higher concentration of sugar, the insect turns promptly and feeds from the richer source. The fly will also turn if contact with food is lost. It may even dance about, as though in a great hurry to get on with the meal. Not until its crop is full of the most sugary solution available does the insect fold its mouthparts and cease its hunting.

Grooming holds great importance for the insect throughout its life. Every speck of dust must be brushed away to keep its minute sense organs exposed. Standing on its middle and hind legs, the fly rubs the front pair together to free them of particles, and then uses them to wipe off its whole head. After rubbing the front legs together an extra time, the insect generally stands on them and the middle pair to brush its hindmost legs against each other, and sometimes along the sides of its abdomen as well. Seldom does the whole operation last as long as the grooming behavior of a house cat. Yet care of the body surface may be even more important to the insect because any speck of dirt is proportionately so much larger. Not including its tail, a cat is about sixty times as long and thousands of times more bulky than a fly a quarter-inch in length.

The larger size of a praying mantis and the habitual pose of
this predatory insect—which stands almost motionless for
minutes at a time while waiting for prey to come within snatching distance—makes its grooming behavior easy to observe. At
frequent intervals the mantis raises one foreleg, then the
other, and swivels its head on a flexible neck while cleaning all
of these parts in elaborate detail. A leg may be used to bend
a flexible antenna, scarcely thicker than a cat's whisker, until
it can be brought between the mantis's jaws. As though to
nibble off any dirt and lick the antenna clean, the insect
grooms its feeler farther and farther toward the free end.
Sometimes, no doubt by mistake, the mantis will clamp its jaws
together and permanently shorten an antenna. At the end of
summer, when the insect is mature, both its antennae may be
too short to groom at all. But its wonderful vision, its snatching ability, its appetite, and its interest in a mate often continue
until the night of the first killing frost.

Because it has fewer sense organs and different nervous
coordination, any insect will be more selective than any mammal in the stimuli to which it will respond. Yet to any suitable
sight, scent, or sound, the insect reacts promptly. Any apparently aimless wandering may merely indicate a scarcity of appropriate cues from the environment. This impresses us particularly when we watch a butterfly flitting in a meadow. The
butterfly picks up olfactory messages mainly through its feet,
as apparently all butterflies and day-flying moths do—differing
thereby from night-flying moths that use scent detectors in
their antennae for guidance. The butterfly need not cease its
sampling of the air while its legs are folded in flight, and the
flying insect tilts this way or that in a breeze. As soon as an
attractive combination of scents wafts upward to the butterfly,
it descends swiftly and directly.

An experience in the rain forests of the South and Central
Amer can tropics showed us how positively butterflies respond
when the lure is right. We longed for a closer view of the
spectacular blue *Morpho* butterflies that occasionally flitted
high overhead as we explored along shady forest paths. These
insects can be seen from low-flying aircraft since their iridescent wings reflect sunlight like flat mirrors. Entomologists who
had spent years in the Amazon basin insisted that the only way

to catch a *Morpho* was to build a ladder up one of the tall trees and stand there, perched high on whatever foothold could be found, while wielding a big net. Seeking a simpler way, we wondered if the butterflies could be lured to overripe bananas. Daily we mashed some of this fruit and exposed it on one of the forest paths. One day, sure enough, the banana odor wafted to a passing *Morpho*. The butterfly circled and descended, settling on the wet mash, which it probed with an uncoiled tongue.

Cautiously, Margery wet the palm and fingers of one hand with the banana mash, and extended this new invitation to the butterfly. At first the insect reacted with fright, fluttering its wings and exposing the metallic blue of their upper surfaces. (Previously it had rested, as *Morpho* butterflies do, with its wings raised over its back so that only the dull brown, speckled undersurfaces were visible.) Then, since we made no further move, the insect crawled upon the wet fingers, flapping its wings gently to assist its ascent. Later, we tried our simple ruse where the light was better. Again the strategy succeeded, and in glorious color, we recorded on movie film a live *Morpho* literally eating out of Margery's hand!

INSECT INHERITANCE
AND PROGRAMS FOR LIFE

The behavior of a butterfly, a moth, a fly, or a beetle can astonish us with its versatility, practicality, and quick adjustments to meet each change in the environmental situation. Every reaction follows an inherited pattern that corresponds to the anatomical and physiological details of the insect and to the sequence of developmental changes through which it attained adulthood. This inheritance is perpetuated with modest variations from one generation to the next and held in coded form by no more than forty thousand different genes—half as many as are found in the nucleus of each human cell. Geneticists know that some genes specify what chemical reactions can take place in the body, others when each process is to occur during the life of the individual. Reproduction at maturity becomes the means for passing along the encoded specifications for the species, making probable its continued survival. Each individual gathers the resources to make more genes, and tests the ones it has against the environment. Sometimes, as a mutation, it tries out slightly altered combinations that might be even more advantageous in its particular environment.

Such differences in inherited gene patterns make possible the existence of more than six hundred thousand separate species of insects, about half of which are beetles. As a result, the insect world now outnumbers in variety all other animals by nearly two to one. Yet the heritage revealed by this vast multitude of six-legged creatures also shows major distinctions that originated far back in time. Only a few have not manifested major changes in body form, behavior, or mode of growth from their ancestors who invaded the land in Devonian times; most notable of these are the bristletails and springtails, which subsist on minute particles of food, grow molt by molt with scarcely a change in body structure, and gain no wings when they finally attain maturity.

A much larger number of insects grow step by step, but may acquire paired wings as their ancestors did in the early millennia of the Carboniferous period. Their wing buds show externally as progressively larger "pads" on the back until, at adulthood, they attain full size. These are the termites, earwigs, leafhoppers, stinkbugs, grasshoppers, and cockroaches. They feed on the same types of food from hatchling to adult and change little in body proportions or behavior until the final adult stage, when reproductive activities appear in their repertoire. The heritage of mayflies and dragonflies goes back about as far, but their feeding habits and body form change rather abruptly when their immature life in fresh water ends and they become aerial adults with expanded wings.

A new and more efficient schedule of life evolved in some members of the insect line just prior to the Mesozoic era (the Age of Reptiles), about 250 million years ago. In these insects the life history of each individual begins with a larval stage, which is committed solely to eating, growing, and surviving. This is followed by a nonfeeding, quiescent, pupal stage, during which the body is reorganized completely into that of an adult with a very different form. We are most familiar with this type of growth in the moth. The moth egg hatches to release a hungry caterpillar; later, the full-grown caterpillar spins a cocoon in which to pupate, and emerges subsequently in final form, as a winged moth ready to mate and perpetuate its kind. The adult always specializes in reproduction. It may also travel on strong wings to new localities where fresh colonies of its kind might become established. It may seek no food itself, but live on the store of fat accumulated during its active larval life.

It seems incredible that for fully three-fifths of the immense time that insects have roamed the earth none of them underwent this complete metamorphosis. Without this transformation, the world would have no beetles, moths or butterflies, two-winged flies, ants, bees, or wasps. Today kinds of insects with a pupal stage outnumber by about six times those that develop by simpler progression, with only minor changes in shape, diet, and behavior.

Transformation (metamorphosis) of a spicebush swallowtail butterfly from the chrysalis. The chrysalides of swallowtails and a few other butterflies are almost upright, held by a silken girdle but anchored at the posterior end.

HANGERS-ON

We might marvel that members of the more ancient orders of insects are able to maintain a place for themselves, despite competition from the vast multitude of more recent kinds. Actually, few orders of insects have ever become extinct; their representatives have always been able to find a distinctive way of life and hang on to their traditions.

Little springtails, for example, who are always less than a quarter-inch long, maintain their wingless existence by eating microscopic plants. Springtails on tide pools along the ocean margin or on the surface of quiet river edges and ponds take advantage of their negligible weight and unwettability to creep or leap about in search of algal cells and pollen grains, fungal spores, and other nourishing particles that fall like dust on the water film. We also find springtails on sap buckets in early spring, when our neighbors collect the liquid from which to make maple syrup. Springtails find their favorite food on snowbanks and are sometimes called "snow fleas." They live high on snowfields and glaciers in the Rocky Mountains, the Andes, and the Himalayas, where the wind renews their supply of food from lower, more temperate elevations. (The only other animals that live so high are small spiders that feed on the springtails.) The world may have only two thousand different kinds of these ancient insects, but they cling to life from sea level to lofty mountains, and from the High Arctic to the hot, humid tropics.

Insects with wings have a slightly shorter fossil history, but some of the most venerable hold to an amphibious lifestyle that probably began before insects of any kind evolved and ventured onto land. Mayflies and dragonflies (or their immediate ancestors) are well represented among the fossils of early Carboniferous times. Their immature stages are invariably aquatic, well fitted for taking what oxygen they need from the small amounts of it dissolved in water. After a mayfly naiad has digested a lifetime supply of food, scraping it from the surface of rocks in a rapid river or burrowing for it in the bottom muck of slower streams and lakes, the insect rises to the water surface and bursts through its outer casing. Then its wings suddenly expand, and the mayfly flits to the nearest cattail or leaf

on an overhanging tree. There the creature clings, waiting for
further changes that will make it an able adult. Even then it will
show no interest in food. It will, however, join an aerial ballet
somewhere in search of a mate. After fertilization, the female
places her eggs where they will hatch into aquatic young.
About fifteen hundred different kinds of mayflies still follow
this ancient routine—if a fish or a bird or a bat does not
interrupt it.

More than three times as many different kinds of dragonflies
and damselflies develop in freshwater shallows as predators on
whatever smaller creatures they can catch. The prey may be a
mayfly naiad or a young fish, both resources that have been
available throughout the extended span of dragonfly existence. If the dragonfly or damselfly naiad does not itself fall
victim to some hungry neighbor under water, the fully-grown
naiad has a chance to emerge into the air. There, in a few
minutes, it can extend and dry the wonderful wings that have
been developing in the pads on its back. With them it can fly
—rather gently if a damselfly, vigorously if a dragonfly. Yet
these insects are also unique in an ancient feature: each of
their four wings is separately controlled. Each wing can be
raised and lowered for flapping flight, or moved forward or
backward, or tilted on its long axis. The insect shows magnificent control of all these movements, and uses them to hover,
to rise or descend, or to propel the slender body ahead or to
the rear, at almost any angle.

Certain in ects have been able to hang on to life by varying
their diet as necessary. The omnivorous cockroach is a prime
example. This is the heritage to which Harvey Sutherland
referred in his *Book of Bugs,* when he commented that ". . . man
himself is but a creature of [the very recent past] compared to
the cockroach, for, from its crevice by the kitchen sink, it can
point its antennae to the coal in the hod and say: 'When that
was being made, my family was already well-established.' "

These hangers-on, or oldest of insect forms, also include all
the musicians of the insect world—the cicadas, the crickets and
katydids and grasshoppers—as well as those favorites of fishermen, the fishflies, dobsonflies, and alderflies. The ancestors of
those flies may well have been the first to evolve a pupal
transformation, back in Permian times (between the Coal Age

and the first millennia of the Age of Reptiles). Their larvae are all aquatic, and are found mostly in stony rapids, where they contribute to the food needs of hungry trout.

Larval dobsonflies grow large—to three or four inches in length—and often induce a fish to bite when nothing else holds comparable appeal. Fishermen give these larvae many names: hellgrammites, hell-divers, hell-devils, conniption bugs, crawlers, arnlies, and dobsons.

Only experience tells us that a hellgrammite will transform into a totally different insect—a dobsonfly. After about thirty-five months of aquatic life, the larva crawls ashore, prepares a pupal cell in moist earth, and fasts while changing rapidly into the adult. It becomes a four-winged insect with a span of more than five inches. But it flaps about at night, often to street lamps, in a most uncoordinated fashion. The front pair of gray wings beat independently of the rear pair, never in synchrony like those of a moth, a butterfly, or a member of the order to which the bees and wasps belong. Scientists regard this mode of flight as primitive, and find further evidence that the dobsonflies and their kin are almost living fossils.

Although six hundred thousand kinds of insects overwhelmingly outnumber our human species, relatively few of them have an adverse impact on human welfare, as long as we do not try to manage the environment solely for our benefit. If so many unthinking insects can perform so well, surely intelligent *Homo sapiens* can learn from them to share the world successfully.

2

Making the Most of the Environment

"WHY ARE THERE SO MANY DIFFERENT KINDS OF INSECTS?" people ask when they discover our interest in six-legged life. "They eat everything!"

"Because there are so many different kinds of things to eat," we answer. Sometimes we add, "and because insects have helped to spread nourishment into so many different environments."

In each environment we discover insects testing, tasting, and putting to good use the abilities their ancestors may have developed under quite different conditions. Often an observation about a local species helps us understand its kin in distant places. Almost four-fifths of the kinds of insects native to one continent have relatives with similar behavior elsewhere. Thus, insect actions that we may have missed in one part of the world may be explained by what we see in another location.

THE DESERT

In the deserts of the American Southwest, "the bug that stands on its head" is a dull black beetle an inch long, with legs of unequal lengths. It walks on the sands late and early in the day, perhaps at night as well. So common, conspicuous, and ungainly is the insect that the ancient Aztecs gave it a name: *pinacatl*. The Spanish changed the word to *pinacate* (pronounced peen-a-cot'-tay) and applied it derisively to anyone they regarded as presumptuous (like the English term "whippersnapper"). But no one ever explained in a convincing way why the beetle stood on its head.

In Libya late one afternoon, we found a relative of the *pinacate*, equally black, walking across the desert. Like the American pinacate, the Libyan insect marched along stiffly, head down, rear end raised at forty-five degrees or more. By then we had concluded that its black color helped capture heat from a low sun, keeping the body warm despite the desert's nightly chill. Greater heat during the day sends the beetle deep into the ground, where it finds a more tolerable temperature. But why that posture?

The pinacate beetle of the American Southwest marches across the desert with its body held at a distinct slope; if disturbed, it stops, and elevates its posterior end still more. Each morning before dawn, dew forms on the beetle's back and runs down the slope to its mouth, giving the beetle a drink before breakfast.

Then we met a beetle of the same general type much farther south, on the sands of the Namibian desert. There it stood, head down, in the faint light before dawn, radiating heat from its black body to the space around it. Moisture from the saturated air condensed on the insect's back. The dewdrops grew, coalesced, and slid down to the beetle's mouth. Here was a copious drink of pure water, available before sunrise. This explained what the pinacate gains from having its hindmost legs so long, its front ones so short. One beetle helped explain another.

The vegetationless and drifting sands of the Namib appear to be the most unlikely home for animals. Yet beetles and other creatures thrive there, nourished by the organic particles that constitute more than a third of the material propelled by the continuous wind. Fog often blows eastward across these dunes between midnight and dawn, although all traces of it vanish within two hours after sunrise. Some of the Namib beetles actively trap the moisture, not on their own bodies, but on ridges of sand. Instead of walking normally over the dunes without making more than faint tracks on the surface, these insects dig trenches a yard or more in length. Two parallel ridges of sand rise on each side of the trench, almost at right angles to the prevailing wind. Moisture condenses on the rough ridges, raising their water content to between two and four times that of the surrounding sand. The beetles then walk back along their fog collectors, imbibing moisture as they go. They make the sand work for them in ways that no one suspected, chiefly because people prefer to explore by day. Before midmorning, the wind dries out and flattens the ridges, and the beetles have gone underground for the hot, dry day. Most visitors are none the wiser.

THE REALM
OF THE WATER SURFACE

Wherever air and water meet, an elastic boundary—the surface film—separates the dry world above from the wet below. This interface provides a two-dimensional realm over which a springtail can creep or leap, a midge or a mosquito can rest gently on its wax-coated, water-repellent feet, and some special spiders can scamper without sinking in. The surface film provides a trap for many a downed flier, such as a winged ant or a small beetle whose legs and undersides can easily be wetted. The same special habitat provides the regular territory for at least two different kinds of well-adapted insects: the water striders that scull about dryshod, and the whirligig beetles that swim at high speed with dry backs and wet undersurfaces.

The striders are known as "pond skaters" to many Canadians, "Jesus bugs" to Texans, and water runners *(Wasserläuffern)* to Germans. They scull, glide, or hop atop the water, normally touching the surface film only where stiff brushes of water-repellent hairs protect their feet and hind legs from the knee joint down. Water-repellent scales over the rest of the body serve the same need of keeping the strider dry. Even the sharp, paired claws characteristic of an insect's feet are situated back from the tip, held safely away from the surface film, which they could puncture.

A close look at the reflection of a strider on the water reveals a perfect image of the underbody. Its slender legs slant gradually to meet their mirror images in flexible dimples where the

A water strider standing on the surface film presses dimples that reflect (and refract) the sunlight. Often these bright spots are more conspicuous than the insect, which stands just above its reflection.

liquid supports the strider's weight. The longest dimension of
each oval dimple lines up with the leg that creates it. Five
dimples is the normal quota, for the front legs ordinarily press
into a single depression. Each dimple refracts the sun's rays
like a rimless lens. It focuses each pattern into a dark shadow
surrounded by a bright rim. Where the water is less than
twelve inches deep above a smooth sandy bottom, the encir-
cled dark spots created as the strider presses on the surface
film may be far more conspicuous than the insect itself. Faith-
fully they record its every move.

The sea surface has its water striders too, all of them normally wingless. We notice these insects on the salt waters of
tidal gutters and lagoons, or among the sprawling roots of
mangroves. Some of these striders ride the ocean currents for
immense distances, finding everything they need for life on or
immediately below the water surface. Its salty flavor means
nothing to them, unless they detect it after a rainstorm at sea
that pushes the insect beneath the surface film. With nothing
to crawl out upon the strider must have difficulty getting top-
side again.

Marine water striders often turn up among the animals
caught in tow nets by oceanographers whenever they sample

*Two striders, the one at the right carrying an ant in its beak while sucking the
juices from this prey, approach the empty (or shed) skin of a water spider floating
on the water surface.*

the life at the surface. Lanna Cheng at the Scripps Institution of Oceanography, University of California at San Diego, tells us that samples collected on dark nights contain far more striders than those taken while the moon shines brightly, or by day. Statistical evidence supports her belief that when there is sufficient light a strider can see the net coming and dodge it, despite the fact that the profile of the net is low and its speed approaches five knots.

Whirligig beetles live most of their adult lives half in and half out of the water. Although capable of both diving and flying, the beetle generally races ahead in dizzy patterns, producing wavelets that are both conspicuous and significant.

Each whirligig is jet-black, with an evenly curved, hard shell that is as slippery as a watermelon seed. The largest of them, slightly more than an inch long, live in torrential mountain streams in Southeast Asia. Those in North America and northern Eurasia measure five-eighths of an inch or less. All these beetles seem capable of prodigious speed in the surface film. For its two different forward speeds, as well as for stopping, the whirligig drags or gently waves its middle and hind legs, which are shaped somewhat like a table tennis racket. The legs beat modestly to drive the whirligig ahead at slow speed, generally in a curve, or sometimes in a full circle, like an inexperienced person in a rowboat. More often the whirligig corrects its course, or overcorrects it, zigzagging along in characteristic fashion.

When a whirligig paddles furiously, it pushes the water surface ahead so vigorously that it can stir up as many as fourteen consecutive wavelets. The wavelets form a pattern that remains stationary with respect to the insect, extending as much as $6\frac{1}{2}$ body lengths in advance of the swimmer. Vance A. Tucker of Dade University has captured the patterns in photographs of whirligigs traveling more than an inch per second; the fastest time he has recorded is sixteen inches per second, or nearly nine-tenths of a mile per hour! The best a racing swimmer can sustain over a 1,500-meter course is not much more than four times the top speed of a whirligig.

By day a whirligig watches where it is going with compound eyes that are divided into lower and upper halves. The lower half observes events in the aquatic realm below the surface,

Jet black whirligig beetles dash about over the water surface, past green flakes of duckweed.

while the upper half watches the world above. The upper part of the eye, like the exposed surface of the insect, repels water as though it were oiled. The lower half of the eye, like the rest of the body, stays wet. The whirligig also maintains awareness through short, three-part antennae that it holds forward horizontally, among the topmost molecules of the water film. There the antennae can detect any bow wavelet that is reflected by an obstacle in its path. Such an echo might indicate a nearby insect helpless in the water film—a meal the whirligig would seize upon at once. A larger obstacle creates a slight upcurve of the water film and lifts the antennae, while an irregular contour might indicate the edge of the stream or pond. Both necessitate a suitable change in course. Obviously the insect requires swift responses, since it identifies these situations with only half an inch to spare before collision.

The reflected wavelets also serve for high-speed maneuvering when dozens or hundreds of whirligigs are cavorting on the same small area of quiet water. The speedsters in these insect regattas maintain no discernible formation, and reveal no noticeable conflicts. Whenever anyone tells us that insects never play, we invite them to see a motion picture of whirligigs awhirl and ask for a more convincing explanation. A replay in slow motion reveals no errors in navigation.

INSECTS AND PLANTS

Far more than we generally realize, insects exert a positive influence on the green plants of the world. Soon after the appearance of insects with efficient larval, pupal, and adult stages, the vegetation of the land and fresh waters underwent a tremendous change. Before that time (during the Mesozoic era) the world had no herbs capable of reproducing and dying in a single year. In fact, it had no lowly plants at all, other than the young growth of woody shrubs and trees, and mosses and lichens, which few insects found nourishing. The trees included giant horsetails, clubmosses, and ferns, as well as primitive seed plants, such as the palmlike cycads with tough evergreen leaves. Most of this vegetation grew tall and let the wind disperse its spores. Today these plants are either extinct or considered by botanists to be relics from the past.

Verne Grant, who directs the university botanic garden at Superior, Arizona, credits the beetles in particular with the behavior that pushed the plant world into a real dilemma. The powerful beetles with their clumsy but well-directed flight and strong biting jaws were well fitted to hunt out and feast on the exposed seeds of cycads and other early types of seed plants. At the time, those seeds ripened on leaf surfaces in the treetops, where the wind deposited pollen (spores), ensuring pollination. In Grant's view, the beetles destroyed so large a proportion of the seed crop that plants had to change, to protect their seeds in some effective way—or face extinction.

The ancestors of today's conifers met this challenge by developing cones. The special leaves that bore their seeds gradually evolved into tight whorls loaded with sticky resin that matured into a hard woody protective cone. No beetle could follow pollen into a young seed cone without getting mired in resin. To chew through the woody cone, the insect had to expend more energy to reach the seeds than it gained from them in nourishment. By shutting out insects, the conifers have held to their ancient ways, still growing as shrubs or trees and depending on the wind to carry their pollen. But their numbers have dwindled, so that now the world has fewer than seven hundred kinds of conifers together with their close kin.

The ancestors of the flowering plants became adapted by taking insects into partnership. First, the leaf that bore the vital seeds folded around them, becoming a protective ovary wall that would later ripen as a fruit. Around this closed reproductive part, the plant could transform leaves into a ring of petals, attractive and distinctive in pattern and color. An insect that learned to recognize the display might be rewarded with free sugar solution in nectaries close to the base of the petals, and with pollen—rich in fats and proteins—as it brushed against the stamens on the way to the nectar. The stamens had also developed from leaves. An insect that learned to seek food in flowers of the same kind might pick up pollen from one flower on its body, which would rub off on the appropriate part of another flower. Reliance on insects as pollinators gave flowering plants another new opportunity. Since a pollen-carrying insect would hunt out a flower close to the ground as readily as one in a treetop, the seed plant no longer had to expend energy and time growing to tree-high dimensions for a share of the inconstant winds. By achieving their necessary growth in a few months or weeks, these plants could avoid entirely the more arduous weather between late summer and spring.

For the first time, plants were able to spread out beyond the swamps and forests. They could rely on insects to follow them onto barren lands that became prairies, up mountain slopes beyond the timberline, even into deserts where rain falls only at long intervals. Today the world has more than two hundred and thirty-five thousand different kinds of flowering plants in all sizes, most of them aided by insects. Comparatively few rely on pollen transport by certain birds and mammals (particularly tropical fruit bats), which evolved cooperative roles with plants less than 50 million years ago.

Each year throughout the terrestrial world, the flowering plants follow a regular sequence in opening their blossoms and attracting insect visitors. We notice this exquisite interaction beginning in earliest spring, when the skunk cabbage pushes up through the last of the snow and attracts scavenging flies that have survived the winter in hibernation. They are attracted by an odor resembling that of rotting flesh. Soon comes the floral display that brightly carpets the woodlands, as hepatica, trillium, anemone, and starflower open wide their petals for insects with short mouthparts. Butterflies with a deep thirst have not yet emerged in numbers, and the long-tongued moths need a few more weeks before they make an appearance. Their time coincides with that of the deep-throated flowers, such as columbine, phlox, and evening prim-rose. Wild carrot (queen anne's lace), asters, and goldenrod wait until almost autumn, when the bees and wasps must gather food to survive the cold months ahead, and butterflies, such as the monarch, need energy to travel thousands of miles on their annual migration to warmer regions.

The most effective guests of flowers, admittedly, are the professionals. Bees, in particular, nourish new generations of their kind on evaporated nectar and half-digested pollen of the plants whose cross-pollination they ensure. Ordinarily each bee tends to concentrate its efforts on one type of flower on each foraging trip, day after active day. Honeybees will laze in the hive or on some leaf nearby until their flower-of-the-day opens. Only when the flowering season for that particular blossom tapers to an end does the bee begin to perfect its approach to a different type, to discover which blossoms will yield the most reward for the least amount of effort.

No bee, of course, calculates as a person might the cost/benefit ratio from increased skill in entering certain flowers. It merely improves its performance by following inherited guidance while responding to stimuli from the outside world. The bee remains active long enough—about two weeks for a honeybee and two months for a bumblebee—to encounter a succession of different plants in bloom, learning to approach each type of blossom by the best route, to open or manipulate its parts, to collect the rewards, and to be off to the next flower.

We notice a whole range of skills each year when the bum-blebees in our home area come to the pink ladyslipper orchids that bloom in the shady pine woods. Experienced bees fly full speed into the flower, disappear through the slit that opens

momentarily in the "slipper," slide down the interior, comb
out some nourishing droplets from the "food hairs" inside,
inadvertently brush pollen on the stigma where it will help the
plant, and then head for one of the two rear exit holes where
light shines into the blossom. Each exit is situated beside a
stamen, which spreads more pollen on the bee's hairy body.

Bernd Heinrich, an entomologist on the faculty of the University of California, keeps close account of bumblebees and
the flowers they visit both around Berkeley and on his summer
farm in Maine. Recently he offered the local bees on his farm
a new experience—blossoms of introduced European monks-
hood *(Aconitum napellus)*, a plant not known to grow in his part
of New England. He screened off the whole patch of monks-
hood until it was in full bloom, then watched carefully as
bumblebees discovered the bonanza he exposed. They investigated unopened buds as well as expanded flowers. Only a
minority discovered how to get at the nectar and pollen, but
those that did learn the trick soon had a success rate of 100
percent, and traveled from one flower to the next, exploiting
the new resource. The bees that failed to learn how to feed
from the flower soon gave up and missed a rich supply of food.
Heinrich noticed that the plants show a definite parsimony in

*Bumblebees warm themselves and visit the earliest flowers—the arctic blossoms
and many others in the Northern Hemisphere—having learned by experience how
to obtain nectar and pollen from each kind.*

attracting pollinators. They offer just enough nectar to keep the particular insects that serve them coming—and only at the time of day when these insects are most active.

In general, a large blossom exposed by itself that requires a bee to make a separate landing and takeoff will be visited only if it offers enough fuel to send the insect pollinator on to the next bloom of the same kind. Smaller, clustered flowers offer less nourishment, because any bee that settles on them can crawl from one blossom to the next expending less energy than if it had to use its wings. Conservation of resources pays off for the plant in this mutual give-and-take with a pollinating insect.

Flowers that open early in the spring and depend on bumblebees have to supply extra fuel to keep the insect warm while its work goes on. Bumblebees are able to forage earlier in the year and over more degrees of temperature and latitude than related insects, thanks to their ability to keep warm by muscular shivering prior to takeoff and, in cool weather, also during foraging expeditions. The bulky body and insulating coat of hairs help considerably as the insect raises the temperature within its thorax until the wing muscles are in the proper operating range—between 85° and 110° F. To make any profit on this investment and enterprise, the bee has to gain more than 0.54 calorie per minute from the flowers it visits. The alternative is to remain idle, letting the body temperature diminish to that of the surrounding air.

Differences on the plant side of this partnership are especially obvious in the cues that attract the insect. Flowers that are deep red or blue cannot be seen after dark, whereas white and off-white blossoms (whether ivory-hued or even yellow) remain visible even under starlight. Night flowers offer their nectar from evening twilight until dawn and attract the long-tongued moths that whirr and flutter with surprising dexterity. Jan Purkinje, a Czech scientist with insomnia, noticed the same difference in his own ability to see flowers. He tested his discovery and distinguished two systems at work in the human eye, one for vision in daylight and bright moonlight, the other for night vision. During the day, his sight was less sensitive than at night, but he could see the red and violet ends of the solar spectrum. At night, he was color blind; red and violet became the equivalents of black. Insects too distinguish only shades of gray after twilight fades.

Most bees will remember for a few days the color, general shape, and odor of any flower in which they find an abundance of sweet nectar. Simple experiments can demonstrate this in any summer garden. We often repeat a procedure shown us by Niels Bolwig while he was teaching in South Africa at the University of Witwatersrand. The first step is to dissolve a little table sugar in water as an imitation nectar, then scent it with a few drops of cologne. Any bee that finds this lure will make repeated trips for more and drink all it can hold, until its flight to the hive appears somewhat labored and follows the most direct route.

To make sure that we are watching the same honeybees, we catch about fifteen as they forage for nectar from real flowers, and daub each one's back with an identifying mark of harmless paint. Then we release the insects one at a time atop a clear glass plate set on a piece of blue paper. Before the bee can fly away, we smear one drop of fragrant sugar water on its head. The bee cleans its head and drinks in the sugar water, ignoring the paint on its back. Five or ten minutes later the marked bees are back for another load. This time they find the same glass plate covering imitation flowers made with five or six petals of the same blue paper. If a "flower" has a drop of cologne (but no sugar water) at its center, the marked bees will settle there without hesitation nine times out of ten. If, instead, we place the scent spots at the tips of the paper petals, they will receive almost all the attention.

Over and over the bees demonstrate that they associate the odor we supply with the sugar, and they remember the blue color they see at departure, even though the blue may represent just a general target where olfactory cues take over. Many flowers secrete odorous materials on their inner surfaces in successive spots like the luminous markers along a runway at an airport. These lead the bees to nectar deep inside the blossom.

Ultraviolet radiation from the sun, although invisible to our eyes, reflects from many flowers in patterns that insects see plainly. The scanning electron microscope has been turned on flower petals and has confirmed that the epidermal cells rise up in minuscule fingerlike papillae. These carry the chemical molecules that create patterns by reflecting or absorbing the ultraviolet in ways that an insect can detect. Even the nectar itself may absorb ultraviolet radiation and emit light by fluorescence in other parts of the spectrum. The intensity of this visible energy provides a fair measure of the amount of nectar available. The sweet solution glows a bright blue in onion flowers and in flowers of the almond tree. Its color ranges from ultramarine to yellow in buckwheat blossoms, which may explain why this lowly herb is such an outstanding honey plant.

Color is surely important to the insect. A worker honeybee that visits a dozen blooms of a single type but of various hues, all in a few minutes' foraging, will concentrate attention on blue flowers to the exclusion of others if it finds three blue ones to which a drop of imitation nectar has been added. If no more blue flowers of the same type can be found for two weeks, the insect may remember its earlier success and land on the next blue flower it encounters.

The shapes of flowers are less significant cues to insects. The number of petals matters less than the proportional area of the flower to its perimeter. Somehow the insect eye measures this ratio and detects the difference when the flower loses a petal as it ages. An old flower has less perimeter and might be ignored, for it probably has neither pollen nor nectar to make investigation worthwhile.

Even the plant world is not above unwitting deceit. Some orchids induce helpful insects to visit their flowers, but offer nothing nourishing in return. Across Eurasia, members of the orchid genus *Ophrys* produce flowers that resemble bees of various kinds and also emit a fragrance that these insects and a few related kinds will recognize and seek out. The scent differs in no detectable way from the odor of the female of the insect species. The male insects come to the *Ophrys* flowers and rub against them in an imitation of a courtship display. From one blossom to another they go, transferring pollen with no lasting reward. The plant thus seduces the insect by mimicry in appearance and scent, revealing no differences that the insect can detect. Other orchids maintain a failsafe system. If no visitor arrives with the behavioral key to unlock their special entry system, to bring pollen and take away more, these blossoms pollinate themselves. About seventy-five different species of orchids, or roughly 3 percent of the known species in this huge family, never open, depending on themselves for reproduction.

HUNGRY INSECTS, EDIBLE PLANTS

Since there are fewer than three hundred thousand kinds of terrestrial and freshwater plants that provide the nourishment for more than six hundred thousand different kinds of insects, an overlap in dietary habits is unavoidable. Perhaps only sixty thousand kinds of insects do not attack plants directly, getting their energy and raw materials for growth from other animals by predatory or parasitic behavior. Fewer types depend on nectar and pollen. Still others live as scavengers, eating little that we regard as "living," although live bacteria must be a considerable fraction of their food.

The clearly vegetarian insects minimize their impact on live plants by being frugal and behaving as specialists. No individual insect overeats, and an abundance of food merely supports more individuals. Each individual gains weight until either the insect reaches full size or its time to change to adulthood, when it transforms its energy reserve into organs for reproduction and dispersal, leaving some as fat for future use.

By developing different ways of extracting nourishment from the same plant, various insects have both lessened the danger to the plant itself and facilitated the cohabitation of species. One example of a plant that nourishes more than one kind of insect is flax, a member of the lily family native to New Zealand. The strong fibers of the plant, which can be used for making cordage, run the full length of each narrow, spear-shaped leaf, and the leaves sometimes reach a length of nine feet. Two kinds of caterpillars attack flax leaves: the flax-notcher, a measuring worm that eats U-shaped or V-shaped notches from the edge of the leaf, and the window-maker, or larva of the owlet moth, that feeds near the strong midrib and cuts elongated openings right through the blade between the veins. A window-maker and a flax-notcher can pass each other half an inch apart during their feeding activities without being aware that the other exists. Neither of them interferes seriously with the ability of the plant to make use of its foliage, or destroys many of the valuable fibers.

Seldom does anyone notice how carefully each hungry caterpillar adjusts its inherited pattern of behavior to suit the opportunities in its environment. We can predict where and approximately when to look for a demonstration by noticing in midwinter the sheaths of eggs surrounding cherry or apple branches, where a mother tent-caterpillar moth has laid them, then coated them with a weatherproof secretion resembling lacquer. In May the eggs will hatch and small caterpillars will begin at once to spin a communal web. This rainproof structure serves as a shelter, a solarium, and a place of privacy for the brothers and sisters, all fifty to three hundred of them, all

The egg mass of a tent caterpillar moth appears varnished, a winter-resistant cuff around a branch.

still too young to be aware of sex. The young caterpillars stay inside every night and on cold and warm days, but only part of the time on cool days. Cool weather signals them to emerge from the few small doorways in their shelter, and to walk in single file toward brighter light, to reach young, soft, edible foliage. A change of weather or the approach of darkness sends the caterpillars back to their nest where they conserve their energy. They also return within ten to thirty minutes after each mass exodus and stay concealed for about six hours before showing themselves again. This inbuilt schedule confines the insects for the hours they need to digest the foliage they have chewed and swallowed. Fecal pellets quickly blacken the inside of the silken tent, repelling birds and mammals that might otherwise tear the nest apart and eat the insects. The caterpillars, however, minimize traffic through their own wastes by adding new decks to the tent. Each extra compartment has about half an inch of headroom through which one individual can easily pass another.

The multiple branching of a tree must present a real challenge to a shortsighted caterpillar. Yet these insects find their way rapidly to fresh foliage and back to the nest. Rarely does a caterpillar explore a branch that has already been defoliated, because it can quickly tell whether others of its kind have traveled a particular route, and whether the most recent of these travelers came back from a leafy feast.

The branch itself shows no wear from the feet of marching caterpillars. It does, however, shine like a carpet runner where they have been. Each insect spins a silken strand as it marches along. A well-traveled trail will bear hundreds of strands. In 1976 T. D. Fitzgerald of the State University of New York at Cortland discovered that a tent caterpillar, returning from a good supply of food, adds a flavor that lasts for hours to its silken trail. The substance alerts any outgoing caterpillar that it is on the path to a feast. Caterpillars returning from a defoliated branch add no such flavor, and their silken tracks are avoided. If a tree is small or bears multiple colonies and is soon denuded, a few caterpillars are likely to lay a silken trail across the ground to another tree and return along that trail to the nest, thereby informing their siblings and cousins of a new food source.

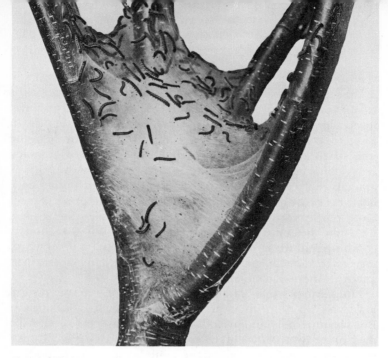

For a while the caterpillars emerge from their tent only to feed briefly several times a day; each one has a tremendous appetite for foliage. By mid-June they have spun cocoons in isolation, in which to transform to moths. Females fly in the autumn and place more eggs in suitable sites. (Photo courtesy of U.S. Department of Agriculture, Bureau of Entomology and Plant Quarantine)

Only when tent caterpillars attain their full length of about two inches do they lose interest in food and in fellowship. Individually they set off to find a sheltered site, such as a generous crevice in the bark of a different tree. There a caterpillar can spin its private cocoon and spend the rest of the summer as a pupa. Autumn will call forth the adults to mate and provide new clusters of eggs to survive the winter.

Many a caterpillar and many an adult moth fail to fulfill this potential destiny. A surprising number of fully grown caterpillars simply stop walking. They stiffen and die of a disease that does not cause an obvious change in behavior or in the form and color of the plump insect prior to death. Other caterpillars may attract the attention of a hungry bird or toad, or be carried off by ants in daylight, or be eaten by skunks and opossums at night. In one way or another the population of tent caterpillars, and of the moths into which some mature, shrinks each year until relatively few survive to perpetuate their kind.

Where trees are scarce or the diversity of tree species is small, as in the northern coniferous forests or on a Tree Farm, insects find fewer opportunities and fewer natural controls. In a very real sense, these regions remain frontiers. In them the web of food relationships has not yet developed complex cross-connections between the green world and the various animals that eat plants or one another. Only such cross-connections provide stability. In the Far North and mountain regions of the Northwest, the glaciers of the Ice Age melted less than ten thousand years ago; neither time nor the climatic conditions in these regions have yet permitted a diversity of insects to develop and contribute different checks and balances to the environment.

Lumbermen prefer single-species stands of trees of uniform age as a crop, since they can be harvested most economically. But so vulnerable an environment invites insects to cause the premature death of useful trees over great areas. Many foresters still cite the outbreak of the Douglas-fir tussock moth in 1946. Its caterpillars killed virtually all of these valuable trees over sixteen thousand acres of forest near Moscow, Idaho. Applications of insecticide from aircraft in 1947 supposedly saved the firs on 384,000 adjacent acres, permitting the harvest of over 1.5 billion board feet of timber valued at almost $85 million.

In modern North America, only four kinds of insects significantly reduce the yield of timber and pulpwood. The Douglas-fir tussock moth, whose caterpillars attack these trees in the Far West, is still one of them. Another is the spruce budworm, which does much damage to potential pulpwood in Minnesota and Maine. A third is the southern pine bark beetle of Georgia and adjacent states. The last is the famous gypsy moth, a Eurasian species that was introduced carelessly in 1868 near Boston and has spread ever since. Even these four, however, attain outbreak numbers chiefly following certain combinations of weather and tree-cutting operations. So far, control seems impossible without greatly increasing the cost of taking out trees for commercial use.

Some foresters now question whether attacks by insects are always bad. Entomologists W. J. Mattson and N. D. Addy of the

U.S. Forest Service tell us that most other insects that eat nutritious foliage and inner parts of trees may actually benefit the forest. All healthy trees produce more leaves than they need for maximum growth and reproduction. Their shade becomes excessive, and their extra leaves store much mineral matter in a form that resists decay for years. Caterpillars and other insects may destroy up to 30 percent of the foliage, and kill many trees that are declining in vigor, without harming the forest as a whole. The insect droppings return to the soil the mineral nutrients the living roots require. Caterpillars open up the forest canopy, letting more warm sunlight and rain reach the ground. The recycling of raw materials goes faster, increasing the vigor of the whole stand of trees. Even the coniferous forests, which long ago shut out hungry insects from their developing seeds, benefit from the recycled mineral nutrients that insects continue to drop close to the roots.

This complex interaction between vegetation and insects developed so gradually, over so many hundreds of millions of years, that its many origins are impossible to trace. Yet we can scarcely imagine any insect going in search of something to eat, or placing an egg where its offspring will find appropriate nourishment, without wondering what stimulus from the food source initiates this behavior. The insect continually seeks stimuli that will prompt it to follow a sequence of inherited responses, and the stimuli are usually present when the insect arrives.

Only a minority of insects actually interfere seriously with human plans, and we should be highly selective in taking drastic measures against the few that seem to be perpetual villains.

INSECT ATTRACTANTS AND REPELLENTS

Every kind of vegetation releases volatile chemical substances that evaporate and are detected by insects at a distance. Some offer odors to which a human nose responds, often with pleasure. Recognition of this fact has led to the establishment of walking trails through public parks where blind visitors can follow a guide rope to small signs in braille and outgrowing branches of plants, which can be felt and sampled for fragrance. How much more significant this is for a hungry insect as it seeks edible vegetation, and how much simpler, because the sampling is done at extremely close range.

Botanists have long wondered why vegetation synthesizes and releases compounds that have no known role in the plant's growth or reproduction. Biochemists have identified many of these substances, and quite generally scientists now assume that such materials benefit a plant by reducing the number of kinds of animals that feed on it. The repellent substances, while effective in driving away most plant-eaters, lure a few species of insects that possess immunity to the particular poisons. This paradox brings to mind an old saying of the Roman poet Titus Lucretius: "What is food to one man may be fierce poison to others."

Knowing about this phenomenon does not make our neighbor admire the small leaf beetles that ravage his cucumber vines every summer. Throngs of them arrive to feast on the foliage. Sometimes they attack the leaf stalks, flower buds, and parts of the open flowers as well. The slightly larger of the two kinds of beetles are pale yellowish-green with a dozen black spots on their backs. The smaller beetles bear a few lengthwise black stripes on backs that range from yellow to brownish-yellow. Farther south, identical beetles attack watermelon stems and leaves. These insects seem peculiarly attracted by (and tolerant of) a class of poisons known as cucurbitacins. Most other animals find them distasteful or toxic; they may even be cancer-inducing for mammals. We enjoy the fruits of these plants only because horticulturalists have largely rid the cultivated varieties of the bitter substances in the edible parts. By doing so, the growers have simultaneously made the fruits unattractive to cucumber beetles and susceptible to damage by other insects.

Horticulturalists at Purdue University in Indiana recently noticed great clusters of cucumber beetles feasting on some particularly bitter watermelons that had been cut open. These fruits were from a mutant strain that seemed to be a biochemical throwback to the original watermelons from Africa, which have long been known to poison any cattle that eat their leaves.

The beetles concentrated on a layer just under the skin, as well as the heart of the bitter melons. There the fruit proved to contain the highest and the second-highest concentrations of the toxic cucurbitacins. The insects ignored the red flesh and the region around the seeds, which had the smallest amount of the substance. Honeybees and wasps ignored the bitter fruit altogether, but came readily to cut surfaces of sweet watermelons—fruit which had no appeal for cucumber beetles. Seemingly the beetles could denature the dangerous compounds, thereby gaining a food resource free of competition.

Ever since the International Centre of Insect Physiology and Ecology opened its laboratories in Nairobi, Kenya, in 1971, biochemists there have been exploring repellents in native vegetation that they call "antifeedants." The researchers aim to identify and perhaps synthesize substances that either prevent damage by insect pests or attract them to traps in large numbers. "Not surprisingly," these scientists reported in

Toxic materials secreted by plants reduce the number of different kinds of insects that feed on them. Milkweed is host to the poison-tolerating caterpillar of the monarch butterfly, the pink longicorn beetle, and, as shown here mating, some red-and-black juice-sucking bugs.

March 1978, "the antifeedants found in this program turned out not to be uniformly active against all insects but appeared to have some . . . specificity."

A plant used extensively in African folk medicine and highly regarded in Kenya for its effectiveness with some diseases proves to contain "one of the most potent antifeedant compounds against armyworms." It is almost without effect on one of the large vagrant kinds of grasshopper and on the tobacco hornworm caterpillars that were introduced inadvertently from North America. These two kinds of insects will feast on the plant, and seek it out if they can reach it. Other compounds that prove active against the introduced Mexican bean beetle have no effect in repelling armyworms. Too often a pest species is attracted by the very substance that repels others, but a whole arsenal of repellents and attractants awaits discovery.

THE OAK TREE
AS A SUPERMARKET

Of all the trees in the Northern Hemisphere, oaks are most likely to impress people with their sturdiness, towering growth, hard wood, and reliable shade. Even the elements cannot soften an oak leaf until many a rain has leached away its tannins. The "old oaken bucket" was equally resistant to decay. Yet the living oak tree is a regular supermarket for insects that can tolerate its repellents, and it houses companions at so many levels that it also resembles an apartment house.

At basement level, deep in the soil, dozens of cicada nymphs extend subterranean passageways to reach fine oak roots and rootlets, from which they obtain liquid food. Each nymph is a robust little creature with front legs specialized for digging, a sharp beak fitted for probing the oak, and developing compound eyes that will one day have a use above ground. These nymphs may take a dozen years or more to complete their development, but the damage they cause individually is not excessive.

At ground level the oak spreads a banquet of acorns among its fallen leaves. Where acorns have fallen only recently, we search for those with a tiny white dot on the shiny surface. It marks the place where a female weevil with a particularly long snout chewed a channel straight into the acorn. After depositing an egg in the cavity, she neatly seals the opening with a bit of excrement. If we cut the acorn carefully, we may find inside it the grublike larva of the acorn weevil, and sometimes a caterpillar or two as well. A single caterpillar may not necessarily prevent an acorn from germinating later, but the developing weevil larva will destroy the seed. Fortunately, the number of these insects is controlled by squirrels, which seek out the infested acorns as an attractive change in diet.

The tender branches of the oak tree attract treehoppers and aphids, which insert slender mouthparts into the tissues and work deeper until they enter a conducting tube in the inner bark. Here the tree transports sugar solutions—the valuable product of photosynthesis—toward growing regions or storage sites. These solutions are under pressure in the tree, and once the insect taps an active sieve tube, nourishment is actually pushed into its body. More sugar solution goes through the insect than it can use, and the excess is discharged as honeydew. If ants do not arrive quickly to lick up the excretion, it drips to the ground. The buffalo treehopper is the one we see most often on oak trees. We recognize it because the forward part of the body is disproportionately large, like a bison's head and shoulders, and twin projections like horns point to the front and sides of its head.

Drops of honeydew will often speckle an automobile parked beneath an oak tree, the sugary liquid drying as hard as taffy. Particularly large amounts drip from the legless and almost spherical females of certain scale insects of the widespread genus *Kermes.* Each of these insects appears pegged in place by her slender mouthparts. Despite her jointed antennae we could easily mistake her for an abnormal growth on the twig. Members of a relative that lives on scrubby oaks in arid parts of the Near East spatter the stiff foliage with honeydew, which dries into flakes, drops off, and is blown away by the wind. The sugary substance is called manna, the "small round thing, as small as the hoar frost on the ground," that the Children of Israel found and ate, as recounted in the sixteenth chapter of Exodus.

Other sucking insects—different kinds of aphids and lace-bugs—as well as various caterpillars and several kinds of beetles, attack oak leaves before they expand, harden, and become fully charged with repellents. After these insects have fed in their assorted ways, the leaf may be more tattered than a weathered flag. As though to counter this conspicuous deterioration of their world, some of the moth caterpillars collaborate by secreting silken strands and enveloping the foliage of a whole branch, or sometimes an entire tree. This covering conceals the damage being done, and provides a shelter for the insects that is almost impervious to birds.

Between May and July, oak foliage often attracts gold bugs, our most glittery scarabs, which easily win any contest for size and magnificence among the beetles with metallic coloring in any forest. This is the insect that caught the eye of Edgar Allen Poe, who chose it for the title role in a mystery story.

Many of the oak leaves have already nourished tiny caterpillars, known as oak-leaf miners because they tunnel through the leaf between its upper and lower epidermis. They devour the tissues where photosynthesis occurs, creating a visible trace of their progress. One common miner produces moderately

The serpentine route taken by a leaf miner, feeding between the upper and lower epidermis of a leaf, shows as a translucent strip from which the green tissue was eaten and a central dark line of wastes.

wide, winding tracks, while another kind creates trumpet-shaped mines. Although the young caterpillars behave differently as they tunnel, as adult moths they will both look and act alike.

The tough bark and hard wood of the oak appeal to beetles, which are generally more slender, as though adapted to escape through tunnels from the deep locations where their larval life is spent. Many of these insects are metallic-colored wood borers, as lustrous as though burnished. Often we see them clinging in the crevices of the rough bark, where shadows help conceal them from nuthatches and woodpeckers. Other oak-boring beetles are the longicorns, named for the amazing length of their antennae. They ordinarily hold to the bark in an exposed position, with antennae spread, alert for danger, and reacting to any approach by either dropping as though dead or suddenly flying away. The grublike larvae of these beetles gnaw inside the tree. Depending on the species, they may need anywhere from a few weeks to several years to attain full growth. A few become beetles so large (almost two inches long) that they can barely squeeze through tunnels more than a quarter-inch in diameter. When their large larvae chew on the wood, the whole tree vibrates. Anyone can hear the crunching sound by pressing an ear against the bark.

Far more minute insects of almost three hundred different kinds induce oaks to provide them with a special home as well as food. Among them are aphids, tiny moths, midges, and wasps that produce chemical substances causing a leaf or a twig to grow abnormally where the insect is located. The plant tissue forms a hollow bulge, with the insect at the center. The bulge is called a gall, and its shape reveals to the initiated what manner of animal is feeding inside. The most obvious galls on oaks project from either above and below the leaves or from the stems.

One type of tiny gall wasp can cause a spherical gall the size of a Ping-Pong ball to form on a leaf. Inside the smooth tan skin of the gall are strong fibers that radiate from the central chamber where the wasp grub takes its nourishment. Another member of the same gall wasp genus induces a similar gall to form on twigs, but the insect occupies a chamber that eventually breaks loose inside the hollow outer skin, like a seed inside

Longicorn beetles have to bend their tremendously long antennae parallel to their cylindrical bodies to pass through tunnels in the wood of a tree; once in the open, they spread their antennae wide.

some strange kind of fruit. A still different gall wasp in Asia Minor causes local oaks to produce vast numbers of hard galls rich in tannic and gallic acids—acids that are valuable commercially in tanning leather and in making inks of outstanding permanence. An extract from the Aleppo gall on Syrian oaks has contributed to the lasting quality of important documents for more than a thousand years. The legal, literary, and educational professions depend on this material. Gall extract is specified in the formulas for inks used by the U.S. Treasury, the Bank of England, and the German Chancellery, to name just a few. Protective substances from the oak are concentrated in the abnormal growths, forming neat packages that local people break off and sell to make a living. Those that they miss release gall wasps, which carry on the tradition.

Old oaks have dead branches with loose bark and perhaps a knothole or two where a limb has broken off. Insects live in

This beetle has a cluster of tiny spider mites on its back, ready to ride along wherever the beetle flies, in order to reach another tree where they can find their favorite foods.

the knotholes or under the loose bark, feeding on fungi that are attacking the oak or preying on smaller insects that eat the fungi. After an oak has started its slow decline, insects in considerable variety cut through the corky inner bark like vandals. Patches of bark loosen from the main trunk, and insects take shelter there from sunlight, dry air, or the cold winds of winter. After an oak tree falls and its wood crumbles under the attack of more fungi and bacteria, still other insects arrive, such as the smaller stag beetles (which rely on this resource in virgin forests) or the strangely shaped caterpillars of the gray goat-moth.

From acorn to decaying giant, the oak shares the energy it derives from sunlight with insects, which, in turn, transform fragments of the tree into food for themselves and return nutrients to the soil that benefit plant life of many kinds.

UNWITTING PARTNERS

How appropriate it seems for an insect that has spent its larval life at the expense of a plant to become useful to the same plant as an adult! The beetle larva that crunches hard wood within a tree emerges into the sun as a longicorn beetle, trim of body and quick in flight, adept at feasting on golden pollen and distributing it from flower to flower. Likewise, the graceful hoverflies, after maggothood in various situations, settle on one blossom after another to sip a little nectar, watch for a potential mate, and unwittingly cross-pollinate the plant. So often the insect is a tolerable guest while young, and a most useful visitor after it reaches maturity.

In the deserts of the American Southwest, anyone who travels while some of the thirty different members of the genus *Yucca* are in flower has an opportunity to witness an ongoing partnership. White moths with a wingspan less than an inch across come flying out of the dark to the white, lilylike blossoms. The insects are all females, already mated and following an urgent pattern of behavior. First, each pregnant female crawls over the sticky stamens of a *Yucca* flower and gathers up a ball of pollen grains with the unusually large palps located on her head. The moth then flies off with her load to another flower, often on another plant. This time she finds the central pistil and stuffs the pollen mass into a deep pocket at the tip. Unwittingly yet deliberately, she accomplishes pollination (frequently cross-pollination, which is ideal for the seeds-to-be). Next the moth turns and lays an egg (or two, or three) atop the pollen mass she has transferred so firmly.

The pollen of *Yucca* is far too sticky to blow in the wind, and unsuited even to adhere like dust to a less forceful insect. The plant depends upon the moth to move pollen from stamen to pistil; otherwise it sets no seeds after the petals drop. The moth depends on the plant to produce a large number of seeds, a few of which will serve as nourishment for the small caterpillers, which hatch from the eggs she leaves, and ensure another generation of white moths.

A *Yucca* plant in bloom can be admired from a distance as well as close up, for it becomes a conspicuous feature of the environment. It may be a woody, branching joshua tree in southern California. It may be a spanish bayonet in Florida, or so glorious a sight in the desert that we appreciate the Native American name for the plant: "God's candle." The *Yucca* is well adapted to frontier conditions, and is an anchor point in a web of food relations as well as a refuge for animals of many kinds. While exploiting the environment according to their tradition, the *Yucca* moths play an integral part in perpetuating the evolving pattern. They are partners in precarious inter-dependence, important components of the environment themselves. Inconspicuously they contribute to the plants' ability to anchor shifting sands, to capture predawn dew, to stabilize the edge of the living world, and to serve as a base for further conquests.

Insects have changed their specific foods many times in the past, but once they change they tend to concentrate on their new competency. Most dietary items that insects relish today have a far shorter history than the insects themselves. Book-worms, for example, scavenged on organic matter of many kinds until the fifteenth century A.D., maturing as moths or beetles with bodies less than three-sixteenths of an inch long. Once Johann Gutenberg invented movable type for printing books, and craftsmen began binding together the finished pages, bookworms discovered that bound volumes offered glue and leather on which their larvae could scavenge—and thereafter books began to fall apart. Newly matured book-worms might pierce the pages while tunneling their way from food to freedom. Now, almost at the end of the twentieth century, librarians at Yale University believe they have a rem-edy. Henceforth, the precious tomes in the Belnecke Rare Book and Manuscript Library are to be wrapped in plastic and then stored for three days in a deep freeze at minus 20° F. This process should kill the bookworms without harming the books themselves, and the plastic covers should prevent other book-worms from burrowing in.

Insects show by their behavior that their tastes and toler-ances far surpass those of any other form of life, thereby in-creasing their chances of survival. Professor C. Thomas Brues, who guided our graduate studies at Harvard University, told us that "it is practically impossible to find any materials capa-ble of furnishing nutriment that have not already become the preferred diet of some member of this versatile group of ani-mals. Corks, cayenne pepper, spinach, kiln-dried lumber, fur coats, cigarettes, even bologna or the Congressional Record form no exception."

3

The Deadly Hunters

CHARLES LAMB PROBABLY HAD THE RIGHT IDEA WHEN HE SUG-
gested that the custom of saying grace before a meal with a
meat course originated "in the hunter-state of man, when
dinners were precarious things, and . . . something more than
a common blessing." All predators live dangerously, because
their supply of preferred food may be threatened by any mis-
fortune to the natural prey, or even to the green plants upon
which the whole pyramid of life depends for captured solar
energy. In addition, the hunter risks physical damage during
its attack as a potential victim tries to escape or defend itself.

On the whole, the strategies that insects use in securing prey
differ little from those of human hunters, except that people
devise and improvise their hunting methods as circumstances
indicate, and insects mostly follow the same patterns used
successfully by their ancestors. Some take advantage of superior power to overwhelm their quarry. Others attain their goal

—a good meal—by a sneak attack, or by preparing a trap
beside which to rest quietly until some victim inadvertently
gets caught.

The oldest technique among insects with carnivorous habits
suits the needs of solitary hunters in streams and ponds.
Stealthy pursuit, quick capture in plierlike jaws, chewing, and
swallowing are accomplished without moving from the site of
success. Most of these predators benefit from the vigor of
youth, usually the immature individuals later maturing as
stoneflies and dobsonflies and their close kin, or caddisflies.
All these young insects prefer places where fresh water tum-
bles over rocks, or quiet pools just downstream from a splash-
ing waterfall. These waters contain an abundance of oxygen
whipped into solution by the turbulence, and allow the animals
to be as active as they need to be. The insects generally hug
the rocks in the rapid stretches or stay close to the bottom in
the pools below, thereby avoiding the main stream and expending less energy in moving about. There they pursue
mayfly naiads, blackfly larvae, and other creatures that frequent such places. As in any hunt, the slow are caught while
the quick get away. What really surprises us is how few stoneflies, dobsonflies and kin, or caddisflies eat anything at all as
adults, after spending their days of development voraciously
using their strong jaws to devour prey and even one another.

CHEWING INSECTS
THAT HUNT IN AIR

Our favorite predators are the dragonflies and damselflies
that patrol in the sunshine over almost any stream or pond. A
few wear colors and patterns as bright as those of any bird.
Structurally, and probably in behavior too, they have changed
little during the past 300 million years. Fossil remains of ancestral dragonflies have been found in coal seams dating from
the Carboniferous period. Throughout this immense span of
time the dragons have possessed enormous eyes, covering
most of the movable head. These keep everything in view at
once, except to the rear, where the insect's own slim body
creates a blind spot.

From early morning until after dusk, dragonflies flit back and forth, hunting insects of the right size, and earning the name "mosquito hawks." They employ their six bristly legs as a catching basket, and skillfully pass each captive forward and upward to the ready mouth. Refueling steadily as it continues the hunt, the insect displays a degree of efficiency in behavior that would be hard to equal, let alone surpass. No swallow or chimney swift can make such quick changes in direction to follow elusive prey. A dragonfly or a damselfly, munching on its latest prey, may settle lightly and harmlessly on your hand if you stand motionless by the water. The dragon would most likely be devouring a mosquito or a midge, while the damsel prefers an aphid. Wings and sometimes legs are discarded as the predator turns the food and cuts it down to size with sharp jaws. The fingerlike maxillae work the food between the predator's flap-shaped upper and lower lips, into the hidden mouth.

When we notice a dragonfly devouring its trophy in this way, we recall James G. Needham, the "dragonfly man" of Cornell University, and his experiments with these beneficial insects. Once he tried to satiate a dragon he held captive by its wings. He offered it one housefly after another, as rapidly as it would accept them. But so efficiently did the digestive tract of the dragonfly simplify and absorb the food, that the patient scientist gave up after several hours. The insect still acted hungry!

When the flying dragon population in a given area is too numerous to find enough unrelated prey to catch, the larger insects correct the imbalance by simply feeding on smaller dragonflies. The largest dragonfly in North America attains a wingspan of $4\frac{1}{2}$ inches, if it has been well nourished while still an aquatic naiad. Such heroic size has probably led to such colorful names as "horse-skinner," "snake-doctor," or even "devil's darning-needle," although the only harm a large dragonfly or its big naiad can cause is to infest a trout or salmon hatchery.

Today's giants among the predatory insects seem small by comparison with those of long ago. A dragonflylike insect was wonderfully preserved in the coal seams at Commentry, in central France. This fossil from more than 350 million years ago has a wingspan of twenty-nine inches, and its body is much longer than that of any modern insect. A dinosaur among insects, *Meganeura monyi* is assigned to an extinct order (the Protodonata). Although its diet cannot be determined with certainty, it probably dined on its smaller relatives and the immature insects and adults of six other orders that could also fly at maturity. Some were like mayflies, others like bugs, but each must have been available as potential prey both under water and in the air. Only one of these early orders of flying insects has continued to hold its own—the modern-day cockroaches.

Dragonflies of many kinds patrol the air for small insect prey, or rest in character-istic poses on vegetation near the water.

The head of a dragonfly, seen close up, seems mostly eyes. The insect captures prey in its bristly legs and transfers the victim to jaws beneath the head.

Beetles were other early arrivals in the insect world. Particularly alert and nimble are the tiger beetles, many of which shine brightly in the sun and respond to any movement as quickly as a candle flame in a breeze—hence their generic name *Cicindela,* the Latin word for a candle flame. A hungry tiger eats its food methodically, using its sharp jaws to pry apart the shell-like covering of its victim. The tiger masticates the soft internal parts and subjects them to some digestion in its mouth before swallowing the minute fragments and nourishing juices. While it feeds the tiger watches for hazards or further opportunities in its vicinity, and will carry off its prey in its jaws if danger demands a quick run or flight.

Tiger beetles seem to skitter on the ground, running on long, slender legs in search of slower insects. Large protruding eyes give the creature excellent vision, and when a bird, squirrel, or person approaches, the tiger flies off, generally to settle a few yards away facing the intruder. Any tiger will do its best to bite with scimitar-shaped jaws if caught and held against a finger.

Other beetles usually seem reluctant to fly away. Almost all have hard backs and tight-fitting wing covers (actually the first pair of wings), which confer a firmness to the adult insect. Few have any scales to rub off or wing fringes to lose, like moths and butterflies. Beetles often tolerate a near approach and have no stinger with which to retaliate if nudged. These features, and the immense heterogeneity of their body forms and dietary ways, make the insects appealing subjects for observation.

Far more deliberate than tiger beetles are the ground beetles, predators that you see running for cover in the garden when exposed to direct sunlight. Ordinarily they search at a leisurely pace for prey in passageways through the soil, under fallen leaves and twigs, and along the rough bark of trees as high as the forest canopy, where caterpillars are numerous. Most of the world's forty thousand different kinds of ground beetles are similar in shape and are carnivorous, which helps in identifying them as members of the family Carabidae, regardless of size. A few ground beetles have become specialists in attacking land snails. These particular kinds have long projecting jaws and a narrow head and shoulder region, and can easily reach into spiral snail shells to eat the contents.

Many ground beetles are incapable of flying, having lost the membranous rear pair of wings that gives most beetles lift and forward motion. But both flightless and airworthy ground beetles run at the same fairly steady pace—slightly more than one body length each second. Running often takes the predator around its victim, letting the beetle bite its prey repeatedly from all sides until no further resistance is visible. Curiously, a ground beetle will be distracted if several potential dinners

A few ground beetles are particularly well adapted for reaching into the shells of land snails to feast on the flesh of the mollusk. Elongated jaws and a narrow head can follow the snail into its shell.

are in sight simultaneously. The predator is likely to immobilize them one after another before settling down to start its meal.

Among the largest, sturdiest, and most voracious of the ground beetles are the caterpillar-hunters of the genus *Calosoma*. Just one of these insects, an inch or more in length, may kill and partly consume four hundred caterpillars during its adult life.

One caterpillar-hunter caught our attention when it marched sideways out of the grass and attacked a six-inch earthworm that lay on the wet sidewalk early one morning after an all-night rain. Like an officer of the humane society bent on putting some suffering animal out of its misery, the beetle came at the worm every half inch or so, all the way around. Wherever it touched the worm with its jaws, the soft body puckered up. Finally, the worm ceased to react, and the caterpillar-hunter ran back into the grass as though it could finally get on with urgent business there.

Recently the legislators and governor of New Hampshire took time to adopt a hunting beetle—the ladybird—as an official insect emblem. If plants had votes, they might applaud this action, and so should we. Ladybird beetles and their young spend most of their active days in plain sight, searching for soft-bodied aphids and immature scale insects to eat. Quickly the little beetles clean the vegetation of these sapsuckers, thereby leaving the plants with more energy to spend on growth and reproduction.

But New Hampshire has no monopoly on ladybirds, for the four thousand members of the beetle family Coccinellidae that are known by this name are well-nigh cosmopolitan. All but a few live as voracious predators on smaller insects. During the Middle Ages, European ladybirds were dedicated to the Virgin Mary as "beetles of Our Lady" because they rid the vineyards of harmful infestations. The familiar nursery rhyme, which children learn without understanding it, reflects concern for the beetles at the end of the grape harvest, when growers in the Old World customarily set fire to the drying vines:

> Ladybird, ladybird, fly away home,
> Your house is on fire, and your children do roam!

Some people add another verse, which tells of a full-grown larva, spinning a mat of golden silk on which it will undergo its pupal transformation:

> Except little Nan, who sits in a pan,
> Weaving gold laces as fast as she can!

In our western states, ladybird beetles get extra credit for hunting out and devouring great numbers of eggs of Colorado potato beetles and of alfalfa weevils, that would otherwise trouble farmers. In California and South Africa, the Australian ladybird *Rhodalia cardinalis* has been imported and liberated by the thousands to control another Australian import, the "cottony cushion" scale, which was introduced accidentally to America in 1868 and quickly became the chief pest in citrus orchards. These ladybirds lay their eggs next to the young of the scale insect in citrus trees, acacias, and other favored plants. Since a female ladybird may lay a thousand eggs and each hatchling devours about three thousand young scale insects before pupating, the benefits to citrus trees are spectacular.

Ladybirds hunt for their victims until the chill of winter reduces the availability of food. Then many of these beneficial insects seek shelter together in crevices or under bark, where they can survive until the return of springtime warmth. If a March gale should rip away their covering, dozens of ladybirds might be revealed clustered cozily side by side. Certain hilltops in California are famous because millions of these insects rendezvous there by December. The ladybirds can be brushed into boxes, kept chilled until spring, and then shipped to buyers who release them in gardens or orchards.

Neither the ladybird nor its larvae have much to fear from insect-eating birds because at every stage of development the insect is generously supplied with foul-tasting substances. Some are in the yellow secretions that an adult ladybird will streak upon a human finger if held too firmly. As a result, a ladybird can behave independently.

The small eyes of a ladybird beetle seem almost hidden beneath the front end of its hemispheric body, and the insect's legs are so short that, when it stands up to walk quickly, the flat undersurface scarcely clears the ground. We wonder how a ladybird can distinguish between its companions, a potential mate, or a prey. It seems unlikely that one orange-red ladybird can detect the presence or absence of black spots on the back of another ladybird, or the blood-red spots on the wing covers of black ladybirds. More than seventy different kinds (species) live in our part of the world, and can be identified by a specialist from their markings.

Since beetles have diversified so spectacularly that their number of species is now approximately equal to the number of species of all other insects known on earth, we might anticipate amazing predatory habits. Actually, relatively few of these insects are hunters and still fewer reveal any special tricks. A similar carnivorous way of life is followed by the adults and larvae of tiger beetles, ground beetles, diving beetles, and whirligigs, which are all members of the same small suborder Adephaga, or the "voracious ones." The spindle-shaped larvae of the water beetles are efficient predators too, although the adults are vegetarians and of a different suborder.

PREDATORS WITH HOLLOW JAWS

The larvae of diving beetles and whirligigs do have a trick of their own in feeding. Rather than chew up and swallow its victim immediately, the larva pierces it with hollow jaws that serve as hypodermic syringes, filling them with a potent digestive fluid. After a while, the pump reverses action, and sucks a ready-prepared meal into the digestive tract. The empty body of the victim, now limp and worthless, settles to the bottom of the pond or stream.

Diving beetle larvae are known as water tigers. They propel their flexible bodies with fringed legs, pursuing victims and capturing all manner of aquatic animals. A large tiger averages thirty captures daily, including slower insects, worms, unwary fish, tadpoles, and young salamanders. We notice them most often when they approach the surface of the pond tail-first, to renew their supply of air.

A water tiger may be fully grown at five or six weeks of age. It uses its powerful jaws to pull itself ashore, where it digs a cavity into soft, moist earth, and walls itself inside. Two weeks later, its pupal transformation complete, it emerges as an oval beetle. The insect reenters the water as soon as its body and solid jaws harden to spend most of the next one to two years hunting in ponds and slow streams, alternating between the search for living prey among the bottom vegetation and brief trips to the surface for a bubble of air.

The adult diving beetle has a special way of staying submerged and active, which it can do for almost half an hour at a time. Coming up for air, the hunter suspends itself upside

The voracious larvae of diving beetles (known as water tigers) hold on to their prey with jaws like ice tongs, while injecting digestive juice and sucking out the liquid contents. This larva has caught and is holding a dragonfly naiad, while reaching up to the water surface with the abdominal tip to take on a fresh supply of air.

down from the surface film at three points: the tips of its hind legs, spread far apart, and the end of its abdomen. Thus positioned, the beetle draws in air between its closed wing covers and the concave upper surface of its abdomen, where its breathing pores are located. With a fresh load of air, the insect submerges slightly, then exhibits a peculiar behavior. It squeezes the captive bubble of air until the rim projects around the rear of its body as a silvery crescent, and away it goes on its next hunting expedition.

Britain's eminent authority on insect physiology, Sir Vincent B. Wigglesworth, explains how the diving beetle manages so long on a single bubble composed of one-fifth oxygen and four-fifths nitrogen. As the submerged insect exhales carbon dioxide, this gas diffuses through the bubble and dissolves into the pond water around it. As the insect depletes the oxygen in the bubble, more of the gas becomes available from the water and keeps the beetle supplied. Slowly, however, the bubble shrinks because the nitrogen is dissolving too. Eventually the small remainder of the bubble no longer protrudes from under the wing covers, and the beetle surfaces for another bubble. If only a human diver could exchange carbon dioxide for oxygen inside a face mask so effortlessly, long stays under water would be equally possible. But the insect, two inches or less in length, has dimensional advantages we cannot hope to match.

The water tigers and larval whirligigs are not unique in the insect world in possessing hollow jaws with which to jab their prey. A similar adaptation serves the immature stages of terrestrial lacewing flies and their close relatives. Several people have suggested that these little predators be called "aphis-lions." Except for their long projecting jaws, they resemble somewhat the larvae of ladybird beetles and behave like them in patrolling plants day and night in search of aphids and young scale insects. But no one would mistake a lacewing adult for a ladybird. The commonest and largest of the lacewings are gauzy creatures of intrinsic beauty, their rather broad wings pastel green with a fine pattern of crossveins, and their bulging eyes shining, pure gold. If a lacewing is attracted to our house lights and gains entry, we catch it gently, pinching the green wings with care, and release it outside. Then we wash our hands well, because in touching it we have picked up the lacewing odor, which only another lacewing would enjoy. The lacewing's flight reminds us that the insect is a descendant of some of the earliest six-footed creatures to develop a pupal stage. Like its kin the dobsonflies, fishflies, and alderflies, the lacewing flaps its front and hind pairs of wings in casual asynchrony. Despite this apparent lack of coordination, the insect can alight where it chooses at least as well as most beetles, including any ladybird.

HUNTERS WITH
SUCKING MOUTHPARTS

Strong jaws are not a prerequisite for predatory habits. A carnivorous insect can capture prey just as effectively with stiff sucking mouthparts, which it thrusts into its victim. Such sucking mouthparts evolved after the development of jaws that could chew.

Two of these sucking hunters, the water bug and backswimmer, must be handled with care, if at all. Both are vigorous and quite capable of defending themselves. The giant water bug can be formidable, for those belonging to large species attain a length of four inches and a wingspan of more than six inches. Smaller kinds among the approximately two hundred species known are seldom less than one inch long at maturity. They keep their mouthparts ready at all times to deliver a poisonous stab. Occasionally one of these bugs attacks the exposed toes of a barefoot wader, hence its nickname "toe-biter."

Normally a giant water bug lurks concealed among soggy leaves at the bottom of a pond or slow stream. Upon seeing even a fair-sized fish, the insect darts forward, propelled by its flattened hind legs and assisted by its middle pair of legs. The front legs fold in a grasping motion to hold the prey while the bug drives in its beak and waits for its venomous saliva to take effect.

Laurel R. Fox tells us, from her studies of California streams and quiet pools, that the first cohort of giant water bugs to hatch each year gradually shrinks in number, chiefly because backswimmers take their toll. Apparently the bulging eyes of each giant bug are able to size up most potential prey, and it will avoid anything it can't handle. But water bug young that hatch from eggs laid during the summer, although numerous, have little chance for survival, owing to both backswimmers and the cannibalism of older waterbugs, which can usually eliminate the young from midsummer eggs within two or three days.

The smaller backswimmer bugs can be seen more easily in ponds and slow streams, where they appear to be the dominant predators almost everywhere in the world. They can be seen readily while a person lolls in a kayak with eyes only

The giant water bug, seen from below, has grasping forelegs, a sharp, powerful beak, and excellent compound eyes of a dark brown color. The insect sometimes flies to light. It swims rapidly, using the flattened, fringed hind legs for propulsion.

inches above the water film. Buoyed at some modest chosen depth just beneath the surface, the insect appears to hang motionless with outstretched hind legs, unless it chooses to propel itself to a new location. The torpedo-shaped body is rarely bigger than that of a honeybee, and often smaller. But the backswimmer stays inverted, its dark (often black) under-surface upturned, its pale (even white) back directed toward the bottom of the pond. The inverted position is dictated by the location of the insect's air supply. Air fills two lengthwise grooves along the ventral surface, protected by paired lines of overarching, water-repellent hairs. At intervals the insect thrusts its abdominal tip through the surface film to inhale a new supply. Often this is accomplished inconspicuously, be-cause the bug is already resting against the underside of the film like an inverted tripod, the ends of its oarlike legs and its abdomen pressing gently upward, alert to every vibration.

The voracious backswimmers prevent many an immature mosquito from attaining adulthood. Whether the mosquito is still a wriggler, swimming upward to the water surface to inhale a fresh breath of air, or a "bullhead" pupa floating just below the water film and breathing through twin snorkels, it is a favored victim for a backswimmer. Even when a mosquito or a midge has succeeded in emerging into the air, using its floating pupal skin as a raft, it remains in danger for a few minutes. Ready to fly away, the new adult tests its hardening wings by buzzing them 100 to 150 times a second, warming up for takeoff. This action ceases abruptly when a backswimmer rises under the unwary prey and stabs it from below with a sudden thrust of its powerful beak. The backswimmer then sucks in its meal. The empty body of the mosquito or midge drifts away across the surface film. Rarely do we give these hunters the credit they deserve for reducing the number of mosquitoes that otherwise would fly off for a blood meal and perhaps transmit a dangerous disease to a human being or other terrestrial animal.

Laurel Fox finds that, each year, just over a third of a backswimmer's nourishment comes from terrestrial insects that fall into the water and become caught in the surface film. Almost a fifth of its diet is aquatic in origin, from small worms and midge larvae to tadpoles and unwary fish. The balance is other sucking insects of the water world—water boatmen (which take minute particles, both plant and animal, from the bottom), water striders (which feed only atop the water film), the young of giant water bugs, and smaller backswimmers. Few backswimmers survive to be threatened by starvation. As the larger individuals in a pond find fewer mosquitoes and less of other food in early summer, they turn on young backswimmers (frequently their own offspring), thereby controlling the size of the population.

Sometimes a backswimmer descends from the surface and actually dives in search of prey, exploring a shallow pond from top to bottom. If disturbed, it will behave differently according to its age group. A juvenile, which lacks wings, will rush to the water surface so quickly that the insect may break through into air, only to tumble back into the water. S. J. R. Birket-Smith of the University of Copenhagen calls this an "alarm ventilation" activity. He notes that young backswimmers with better control stop at the surface film and spread out the long protective bristles that ordinarily conceal their abdominal grooves. These glistening white chambers fill with air, and each young swimmer can then dive again. An adult, although winged and able to fly, swims away from disturbance or clings immovably to some object, such as a stone or a plant near the bottom. It will even mate under water. It flies mainly to find another pond when, in late summer, its own hunting territory dries up, or to locate a suitable pile of fallen leaves below which to spend the winter.

We wonder why these adult backswimmers bother to move or hide at all. Each is quite capable of defending itself in a way that makes us appreciate its German name, *Wasserwanz* ("water wasp"), or its English counterpart, "water bee." If handled carelessly or brushed against under water vigorously, the insect retaliates. It drives its somewhat conical beak through human skin and delivers a charge of saliva. Pain is immediate and infection may follow, causing further discomfort and swelling, often attributed to venom. Actually, this is neither a sting nor a bite, but a stab.

Occasionally a backswimmer and a water strider arrive almost simultaneously to take nourishment from another insect that has fallen into the water. Invariably the backswimmer gets the prize, perhaps pulling the victim downward into the pond beyond reach of the strider. Even though the backswimmer may seem to be the mirror image of the strider—both insects with ventral surface toward the water film—the adult backswimmer is the stronger and sturdier of the two.

The behavior of water striders as they hunt for prey seems much more obvious than that of backswimmers to any onlooker. But close attention is needed to see how the insect relies on the surface film as a telegraph system as well as a support. Special sensors in each of the strider's long, slender legs inform the insect of any disturbance in its vicinity. Rod K. Murphey investigated this sensitivity while he was a graduate student at the University of Oregon, by studying the comparatively large and common water strider *Gerris remigis.* The insect responded normally to vibrations of the surface film despite confinement in an aquarium adjacent to measuring equipment and under the glass eye of a movie camera. Murphey learned that the sensors for vibration are located in the flexible membrane between segments of the strider's feet. The sequence with which these sensors detect an oncoming ripple informs the insect whether the disturbance is to right or left, front or rear, and almost exactly how much to turn to face a potential target.

If a vibration in the surface comes from somewhere in front of the strider, one quick asymmetrical movement of its middle legs suffices to adjust its direction. If the vibration comes from behind, the strider moves its middle legs in opposite directions, just like an oarsman would turn a boat. The hind leg on the side toward which the strider turns acts as a pivot, and swings slightly forward. The other hind leg, like the front legs, serves only for stable support. The strider sculls straight for its target, pausing about a fifth of a second between each stroke of its middle legs to check oncoming wavelets and correct its course.

A strider of this same kind in New England demonstrated to us its sensitivity and speed as part of a miniature drama that began when a sparrow, hopping across an area of short grass beside a pond, disturbed the guardian ants of a subterranean nest. The bird suddenly began to scratch itself in various places: ants were biting or stinging it, or both. The sparrow hopped a short distance, shook itself vigorously, and flew right over us. An ant that had held on too long dropped into the water a few feet from shore. The creature could not swim, but its struggles set up vibrations in the surface film. From five or six inches away, two water striders responded, quickly turning and sculling toward the ant at top speed. One strider reached the victim and drove its sucking mouthparts into the hapless ant as neatly as any park cleaner might skewer a gum wrapper with a sharp-pointed tool. Off went the strider, ahead of its competitor, carrying its impaled reward. The mouthparts of the strider pumped an anesthetic, digestive saliva into the body of the ant, and in less than three minutes its nutritious contents had liquified; the strider then sucked up its meal and left the empty shell of the ant floating on the surface film. A litterbug!

Several striders sometimes share a larger victim, such as a

winged queen of the big carpenter ant. Their combined salivary output probably helps subdue the prey. Most of the time,
however, the striders feed on smaller insects and do so with
a minimum of commotion. This explains why so many striders
can skate around casually in a very confined area without drawing each other's attention. If one nudges a second, the two
generally separate immediately. Each meeting of two striders
appears to be a mistake, perhaps a misinterpretation of some
vibration in the water. Seldom do the shorter moves reveal
their true significance: they often occur when a strider detects
some small denizen of the underwater realm approaching the
surface film to get air. The strider stabs downward through the
water film, spears the victim, sucks it dry, and casts it away,
without any part of the prey emerging into view.

Hundreds of striders often congregate at midday in the
shade of a big tree. By leaning against an overhanging trunk,
we can see what they are doing without frightening them.
From studies of smaller striders in West Germany, we know
that these creatures can see anything as large as a white face
at a ten-foot distance against a dark background, or a dark area
of corresponding size against the sky. Apparently striders use
their eyes to warn them of danger and to supplement the
perceptions they receive from vibrations sensed through their
legs. Some combination helps them choose whether to approach a new object in their vicinity or hasten away from it.

With the combined efforts of backswimmers hunting under
water, whirligig beetles hunting half in and half out of it, and
striders walking dry-shod on the surface, the boundary between wet and dry can be a busy place. Sometimes it fairly pops
with life because so many other insects rise to the surface, split
lengthwise along the back of their loosened covering, and
escape as adults into air. For a minute or more, each delicate
individual may stand on the water film, allowing its wings to
expand fully, to dry and stiffen. If no predator captures it
during these moments of vulnerability, the creature flies up
into the aerial realm patrolled by dragonflies and swallows by
day and bats at night. The statistics are all against survival.
Only a fraction of one percent may escape with a chance to
perpetuate their kind.

PREDATORS BY PROXY

Before swallows or bats appeared on earth, insects had evolved a new and unique way to be carnivorous, by turning the whole system of predation inside out. Instead of pursuing prey to satisfy their own hunger, adult insects provide edible flesh to nourish their still-unhatched young. These meat-eaters are smaller than their victims, which they devour slowly, selectively but devastatingly, from within. This mode of exploitation is so peculiar and so different from outright parasitism that O. M. Reuter of the University of Helsinki felt a need to distinguish it with a special word. He called "parasitoids" all insects that, as larvae, devour and kill victims provided live for them by their hunting mothers.

This extraordinary arrangement began in Jurassic times, midway through the Age of Reptiles. It provided a new way for larvae to get nourishment rapidly, then undergo a pupal transformation and emerge as provisioning (rather than necessarily carnivorous) adults. At first this pattern of behavior and life history was known only among the kin of gall wasps, for true wasps and bees, like ants, had not yet appeared on the scene. In those days the only insects with a pupal stage were the ancestors of the dobsonflies and their kin, beetles, and some of the more primitive flies (ancestors of midges, mosquitoes, and craneflies). Later, the parasitoid way of life was taken up by some of the more robust flies (the tachinids) and a number of different solitary wasps.

The behavior of these larger, later kinds of wasps intrigues anyone who takes time to watch them. Some hunt only spiders. Others capture grasshoppers, mole crickets, cicadas, leafhoppers, caterpillars, even adult insects as diverse as weevils, horseflies, and queen ants with wings. Each prey requires a specialized approach, a touch (or more) with the wasp's stinger to end resistance, and a particular mode of transport. We see the female wasp carry off her victim, and think of her as a predator. Actually she may not feed on her catch at all. Or, as she hauls away an immobilized caterpillar in her sharp jaws and its limp body oozes a few droplets of its life blood, she may imbibe this food. She does so without significantly diminishing the amount of nourishment remaining in the bulk of the victim, which will be cached to feed her offspring. Otherwise the adult sustains herself on nectar and other plant juices, just as the males of her species do. It is her larvae that will behave as predators on this food supply, or feed from it as parasites. The wasp larva is not too different from that of an ichneumonfly hatching from an egg laid on the host by the female ichneumon. Indeed, the ancestry of hunting wasps is traced to a branch line from that leading to ichneumonflies.

The choice of a victim by a hunting wasp is not determined by special food requirements of the larva, but rather by the limited abilities of the adult female, who must detect, subdue, and transport her victims. The French naturalist Jean Henri Fabre discovered this by experimenting with a sand wasp *(Bembix)*, which not only brings flies of modest size as larval food, but continues to supply them as the larva grows. Since the adult female does not seal off the nursery cell and abandon it, Fabre could take her place. Instead of flies, he supplied the hatchling larva with freshly killed grasshoppers. The larva grew normally on the substitute diet. Subsequent researchers performed similar tests on wasps that stock a chamber for each egg, and let development continue in private. Larvae that normally find a supply of anesthetized spiders can grow just as well if furnished with caterpillars, or vice versa.

Howard E. Evans at Harvard University and other entomologists with a special interest in the behavior of wasps have recognized an intriguing correspondence between the structural features of the wasp body and the behavior the insect shows in procuring and sequestering victims for the larvae to eat. Least specialized are the spider hunters, which first find the spider, immobilize and conceal it, then construct a burrow or a mud chamber; later the wasp returns for her victim, carries it to the prepared cell, and lays one fertile egg on it. The process is repeated with only one spider and one chamber at a time. Only a few spider hunters are like the familiar mud-daubers of North America and Eurasia in building one mud cell after another in the same general location. Fabre marveled that these wasps pay no heed if the cell they have most recently completed is broken open and its contents all removed. Even if every chamber in the cluster is vandalized, the wasp completes her schedule of activities by plastering mud irregularly over the whole mass, as if every cell remained intact.

Sand wasps prepare their chamber in advance, then go hunting for victims. Cells are sometimes excavated in close proximity, showing that the insect remembers where her earlier work was done. Not until she has filled and closed each one in sequence does the sand wasp attend to the general site. Then she either runs about, distributing the sand she has excavated until the entire area is roughly level, or she digs a series of distracting burrow openings, to serve as decoys for other, unrelated insects to find and explore. If they are seeking to find her stored food and larvae, they labor in vain. Both predators and parasites will tire from fruitless exertions and abandon the area of the nests, perhaps without finding a single one.

A third group in the same family of solitary wasps prepares a chamber and lays an egg in it before hunting for victims to feed the hatchling. This behavior is only a few steps away from that of social insects, such as honeybees and hornets. One difference, from block provisioning (walling off the entire food supply for one larva in its own chamber) to staggered provisioning (such as Fabre experimented with in French sand wasps), has already been incorporated into the behavior pattern of a few kinds among the solitary wasps.

Sand wasps prepare their nest burrows before going on a hunt for suitable flies as food for sand wasp hatchlings, one in each burrow. The mother wasp continues to bring more flies as her larvae grow, in a behavior known as "progressive provisioning."

Professor Niko Tinbergen of Oxford University tells of the common European wasp known as the "bee wolf." This insect (*Philanthus triangulum*) loiters near flowers that honeybees will visit. When a bee appears, the wasp hovers a short distance downwind, where olfactory cues can trigger the next move. If the odor is right, the wasp pounces on the bee and grasps it with all six legs. If the tactile cues are right, the wasp stings the victim and hauls the bee to her nest. Sight, scent, and touch act on the bee wolf's nervous system, each feature checked against an inherited measure that triggers the next step.

For many years the hunting wasps were credited with inherent knowledge of the internal anatomy of their prey. Why else would the wasp reach so far under her victim and thrust her stinger into it at such specific points? Supposedly the stinger injected venom into the ganglia of nervous tissue along the midline just inside the victim's body wall, which would have explained the rapid paralysis of the legs and the wings (if any). Prevented from escaping, the creature stays alive and resistant to decay, until the time the larval wasp chews through some vital organ. Now Werner Rathmayer at the University of Munich finds that the bee wolf's stinger rarely penetrates a ganglion. He suspects that the wasp reaches under her victim because there she can thrust the sharp point of her stinger through flexible membranes more easily than elsewhere on the victim's body. Coincidentally, her poison diffuses rapidly from this location into muscles that the victim would use in trying to escape, namely those that control leg and wing movement.

The array of behavior patterns shown by mothers-to-be in this family of hunting wasps may well have developed from less elaborate habits such as are found among the related ichneumonflies and kin. An ichneumonfly shows special skill in finding and attaching an egg to a caterpillar or other victim that is to serve as the food supply for the ichneumonfly larvae. She may affix the egg outside or use her slender ovipositor to inject the egg where the hatchling larva will be safe and can start feeding with no delay. A tachinid fly operates in much the same way. Each species of mother-to-be has her own preferred choice in victims to be used by her young.

The presence of the parasitoid may not be evident until later, however. Moth collectors, who bring cocoons indoors as a way of obtaining freshly emerged and perfect specimens, feel frustrated when a parasitoid insect comes out instead. But, as a result, records accumulate that help us reconstruct the behavior of the female ichneumonfly, even though it progressed unseen by human eyes. For the ichneumonfly that lays an egg into a cecropia caterpillar, the related polyphemus, promethia, and cynthia moths would also make suitable victims, each converting the chemical ingredients of foliage into a nourishing organic form.

We discovered quite accidentally that parasitoid insects do err fatally at times. One March, while tramping through a snowy field, we noticed an abundance of galls on the dead

The largest of the ichneumonflies, with a body more than an inch long, stands on a tree within which a suitable host can be heard tunneling, and slowly thrusts her extraordinarily long ovipositor deep into the wood, then slides a slender egg right into the tunnel where the borer is working. The ichneumonfly hatchling attacks the borer, at first eating only tissues that do not interfere with the continued activities of the host, but later killing the host—usually after an emergence tunnel has been cut through all but the outermost layer of bark.

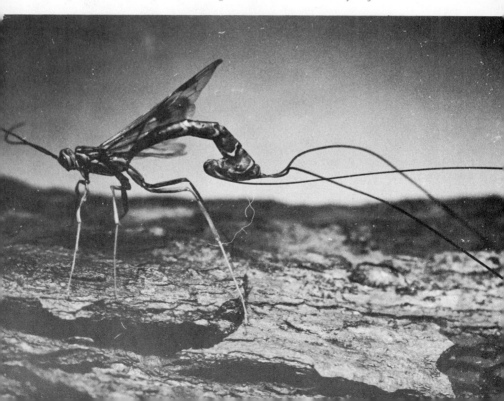

projecting stalks of goldenrods. We broke off more than two
hundred of them, and brought them into the warmth of the
house to learn how many would release adult insects. The
swellings are made by two-winged flies (unrelated to ichneumonflies) that normally emerge between late April and June,
mate, and seek out young goldenrods. A fly egg, laid on a
goldenrod stem, hatches, and the emerging maggot eats its
way into the central mass of soft cells. There the maggot
stimulates the plant stem chemically, inducing continued production of the peculiar swelling. The gall tissue protects the
fly maggot from most hazards, and provides nourishment, too.
Before pupating within the central chamber of the gall, the
insect's final act should be to eat a secret passageway for escape—to (but not through) the thin epidermis of the stem.

In due time flies emerged from most of the galls. We waited

*The green, almost spherical gall on a goldenrod stalk conceals and protects the
maggot of the gallfly developing inside a central chamber. The maggot pupates
in autumn, overwinters in the gall, and emerges in late spring; this timetable
gives pregnant female flies new goldenrods on which to lay eggs.*

two more weeks, and then opened the galls that had no emergence hole. One fly had failed to escape because, as a larva, it left too thick a layer of plant tissue at the end of the tunnel. Two others did not prepare any tunnel, and died in the central chamber—flies with nowhere to go, with genes that would not be multiplied. But several galls contained small wasps. Had they killed the gallfly maggots before emergence tunnels had been completed? Or did the mother ichneumonfly choose the wrong place to leave her eggs, on or beside the wrong maggot altogether? We could not find the answer that year.

The future of any parasitoid seems so precarious, so prone to failures from a multitude of causes, that some experimentation by the mothers may be advantageous. It may locate substitute kinds of victims. We rationalize this way when a starving dog flea bites *us,* for lack of a dog in the vicinity.

SPECIALISTS IN AMBUSH

Ghosts haunt many a boathouse wall or the sides of a dock piling where small craft float on fresh water. Each shows where a stealthy stalker of small insects, tadpoles, and minnows in the underwater world forced its dozen sharp claws into the crevices of the wood, filled its body with as much air as it could, and split down the back to release a dragonfly from its ghostly skin. In an hour or less the broad, shimmering wings of the dragon hardened, and the flying creature darted away to a new life filled with dizzying pursuits.

Until this quick transformation, the dragonfly was an imma-

"Ghosts" of dragonfly naiads clinging to the walls and roof of a boathouse beside a lake show where the dragonflies emerged as flying predators, with totally different behavior patterns from their previously aquatic stages.

ture naiad following a very different pattern of behavior. It crept along the bottom of the slow stream or pond, or clambered in slow motion over some submerged plant. Or it increased the vigor of its breathing movements, drawing in and expelling fresh water through the rear opening of its capacious rectal chamber. Jet by jet, the creature propelled itself forward with scarcely a movement of its legs, barely faster than it would by drifting, yet straight toward its next meal.

The naiad has proportionally longer legs and a broader, shorter abdomen than the parent, but an equal reliance on vision. The creature watches and waits until it sees some potential victim come within its vicinity. Gradually the naiad turns to face this possible prey. Aided by binocular vision, the immature dragonfly can gain a measure of this target—too big for attack, too small to be a meal, or just right. The naiad reacts only when the suitable victim becomes visible in parts of its bulging compound eyes that converge at appropriate angles—in striking range. Abruptly the insect flips forward a grotesque underlip, which springs open, forming a divided scoop. The lip grasps the victim faster than the human eye can follow and drags it back to the sharp jaws, one on each side of the naiad's mouth. Then the lip is folded again, like a mask hiding all parts of the head lower than the compound eyes. Behind it, the immature dragonfly devours its meal and continues to stand motionless.

The terrestrial world also has insects that remain in a single pose for hours, but ready in a split second to snatch prey of appropriate kinds. These are the praying mantises, named for the prayerful attitude they assume, with forelegs held together, until they can be extended to seize a victim. Although distant kin of cockroaches, the mantises stand in plain sight by day, holding their place or swaying gently on long middle and hind legs. The thorax is greatly elongated between the middle legs and the head, and the insect seems to rear up continuously while its triangular head with large compound eyes is well positioned for seeing in most directions.

A mantis will devour insects of many types, the majority of which had not even evolved when mantis ancestors first appeared on the scene. Flies, caterpillars, butterflies, wasps, and bees all came much later, as did flowers of all types. Many

By day the compound eyes of a praying mantis are leaf-green, as is the body. This insect is perched on a goldenrod, watching for some flying insect to come within snatching distance. At night the eyes become chocolate-brown, and more sensitive to faint illumination.

millions of years elapsed before a mantis could stand, as it so often does today, on a goldenrod head and choose which visitor to ignore and which to eat. Apparently a mayfly settling on a tree fern leaf was just as acceptable as a fly or a caterpillar today. The predator can turn ever so slowly to face its prey, and even jump as it snatches to make meaningful contact. Physiologists are astonished by the speed of these actions, for a mantis that misses a fly can make a second try before its prospective victim is airborne and away.

Ambush from Concealment

A New England fisherman who drags a sled house over the frozen bay, from which to catch smelt through a hole in the ice, can appreciate the strategy the larva of a tiger beetle uses in ambushing prey beside a path. Often the mother selects the site, using her jaws to create a pit in which to drop her egg. The hatchling enlarges the cavity, making it a cylindrical pit like a dug well, scarcely larger in diameter than its own body. The larva wedges itself just below the doorway, its feet against one side, two humps on its back and its tail tip against the other. The flat top of its head is level with the ground, although its two jaws project slightly, held apart ready for action. Only the jaws are visible. Sometimes they twitch a little, as though with anticipation, when the larva senses the footsteps of some other insect walking close by. Small eyes watch for the best moment for the tiger to jump up and seize its victim. It is also ready to capture anything that accidentally steps on its head. The ambushed prey gets dragged down into the burrow, out of sight, where the earth protects the aggressor while it makes its meal. The pit serves also as a shelter when the full-grown larva pupates, later to emerge as a young adult tiger beetle.

A far more elaborate pit is excavated by an unrelated but equally grotesque larva with a strangely flattened head. No one would guess that this denizen of sandy soils would transform during a pupal stage into an insect resembling a damselfly, with long, narrow, transparent wings. True to its heritage, the insect flaps its wings in independent pairs. No common name has been given this adult, which sometimes is attracted to lights at night and to flowers by day. Its knobbed antennae distinguish it from any other creature of similar appearance. Its voracious larva—the pit maker—is the only famous stage of the life history, and people call this larva an antlion or a doodlebug.

The mated females of this type of insect scatter eggs in sandy places, where hungry larvae emerge in reasonable solitude. Some (but not all) promptly begin marching backward, digging the tip of their pointed abdomens into the sand; at the same time each larva tosses sand in all directions with its flat, oversized head. By traveling in a circle of decreasing radius, the insect creates a conical depression. It may be nearly two inches deep, and somewhat more than three inches in diameter. The larva lies in ambush, effectively concealed, at the bottom in the center of this trap for ants. Just the sickle-shaped jaws protrude.

A prospecting ant is almost sure to blunder over the rim of the pit and start a small avalanche of sand. The antlion responds by tossing more sand from its low position. The ant loses its footing, slips to the bottom, and is seized by the antlion's jaws. Stiff bristles on the body of the antlion help it keep its balance while it shakes and beats its prey into submission. Like the related aphislions, the antlion has channels for liquid in its jaws. It uses these to suck out the juices of its victim, as though through twin soda straws. One final gesture —a vigorous flip of the head—tosses the empty body of the ant out of the sand trap, like a discarded can of soda.

Insects That Lurk Beside Nets and Snares

Almost any freshwater fisherman is familiar with the billowing nets that certain insects attach to stones and dams over which river water flows. The fine mesh is sometimes an inch across and half an inch high, anchored at both ends and along the bottom. It serves as a strainer for small prey, while the net-maker waits inconspicuously to one side out of sight of hungry fish or birds, until the web catches some food. Then the insect—a caddisworm—hurries over to seize its dinner before the net breaks or the captive manages to escape.

The caddisworm itself resembles a caterpillar, usually with

The adult of the antlion has no common name and little appetite. It visits flowers and meets a mate, and the female scatters her eggs on sandy soil where the hatchlings can prepare pits and be antlions.

firm brown armor plates on head and thorax. Its parent resembles a small moth with hairs instead of scales on the wings, and no coiled tongue through which to sip food of any kind. The commonest kinds belong to the big genus *Hydropsyche,* which comes from the Greek words for water butterfly. The nearest to a common name for the larval stage is "hydropsychid."

The insect follows the same construction method at any site. We find it easiest to watch at the end of a dam, where one can lie face down with eyes just a few inches above the water. Look for a caddisworm holding fast to the dam with a pair of hooked legs at the rear of its body, which will be stretched downstream, tugged by the current. Watch for deft movements of the six pointed legs behind the insect's head as they swing the body slowly in an arc, while the head makes bobbing movements. Every time the mouth comes down to the dam surface, a spot of waterproof salivary glue appears and sticks. As the head rises and swings up and over, it spins out a strand of silk to the next spot of glue. A whole row of these half loops starts the web and holds it to the support. The next row is a trifle wider and each new half loop is cemented to the midpoint of the appropriate half loop in the preceding row. Continuing in this way, the caddisworm creates a strong mesh without a knot or a tear. The current keeps the webbing taut and brings a succession of small creatures where the net will hold them for a few seconds. This is all the hungry caddisworm needs to pick up its meals, rushing from ambush to get one victim after another. It eats enough to last the rest of its life, for, after going through its pupal transformation and emerging into air, it devotes itself to limited travel, resting, and perpetuating its kind.

Net-making caddisworms can be found virtually anywhere swift rivers tumble over rocks toward the sea. But the snare-making fly maggots, known locally as "glowworms" because they are luminous in darkness, live only in New Zealand, where they are prized as a tourist attraction. They are totally unrelated to glowworms (beetle larvae that may transform into fireflies) elsewhere in the world. The New Zealand insect matures as a particularly large, but nonluminous, fungus gnat. As an adult, the gnat eats nothing, but its larvae lure small flying insects that are active in darkness.

Like so many visitors to New Zealand, we went out of our way to visit the famous glowworms that inhabit Waitomo Caves. There, we rode in a small boat on an underground stream to a natural chamber with a vaulted roof. After the electric lights were turned out, and our eyes adjusted to the dark, we saw constellations of glowing insects just under the roof. Each glowworm was suspended in a silken hammock made slippery with clear secretion.

Elsewhere in the cave we saw the same kind of insect at close range. Every one of them had spun from the underside of its hammock a curtain of fine pendant threads. These are beaded at close intervals with clear drops of a sticky material that does not readily dry out. In Waitomo Caves, where no wind can entangle the strands, the curtain of parallel threads may be two inches wide (the length of the hammock) and four inches long. Seen through a large lens, the array suggests the glittering bead curtain in an old-fashioned home, providing some privacy without interfering with circulation of air or people. But this curtain is for small insects that fly in darkness, blunder in, and get caught. The glowworm then hauls up the strand or two that hold a victim, devours the nourishment in the body, drops the empty shell, and spins replacement lines. The glow of the insect seems no more than a come-on, perhaps imitating light from the sky through an opening to freedom.

Millions of these larvae hang their sticky curtains in close proximity throughout the caverns at Waitomo. Small numbers

New Zealand "glowworms" are the larvae of special fungus gnats. Each prepares a suspended hammock on the roof of some cave or cavity through which small insects may fly, then produces long pendant threads beaded with adhesive droplets to catch a meal. (Photo courtesy New Zealand Publicity Office)

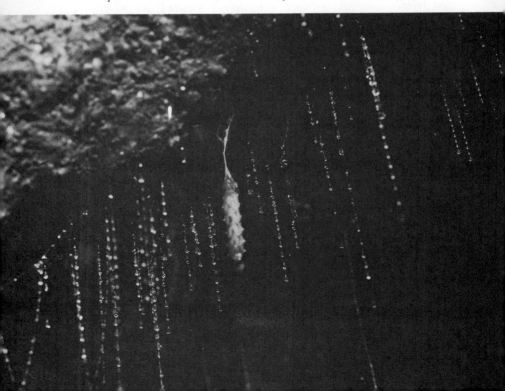

of the same species hang lines no more than half an inch long, we found, in the restricted space under the roots of upturned trees and in fist-sized cavities cut by erosion from the side slopes of steep hills. There the snare-spinners light the way only at night. Their chief requirement seems to be protection from breezes and dry air, with access on one side to a traffic lane through which tiny insects flit in darkness.

Assassins in Danger

Ordinarily, we concentrate attention on the insect that waits in ambush for the arrival of unsuspecting prey. We seldom suspect that the hungry, waiting insect is itself in danger of being eaten by some larger carnivore, perhaps a bird or lizard. Yet enough ambushers and would-be assassins have ways to hide themselves that we should not regard them as the customary end of the food chain.

Recently, a scientific team from Cornell University discovered in the vicinity of Albany, New York, an aphislion that not only dines exclusively on the wooly aphid of alder shrubs, but manages to mimic its victims. The predatory larva snuggles into a close cluster of aphids, feeding on one after another. It salvages the fluffy, woollike wax that the aphids secrete as a covering for their bodies and tangles the conspicuous white material into hooked hairs on its own back. Like the proverbial wolf in sheep's clothing, the larva imitates its victims so well that the human eye has difficulty distinguishing it. Nor does a combination of vision, scent, and touch allow detection of the ambusher by the ants that tend the aphids for their honeydew. If an aphislion is deprived of its wooly disguise, the ants quickly seize the invader. Generally, the larva is too tough for the ant's jaws to pierce. The ant, however, will carry away the ambusher and drop it to the ground. Some other predator is likely to eat the naked larva before it can return to the aphid cluster and renew its protective cover.

Other aphislions—as well as the larvae of lacewing flies—construct shelters from the emptied bodies of their own victims. These too adhere to the hooked hairs on the back, letting the ambusher move about as a sort of traveling morgue.

The same strategy, using the remains of prey, is used by a

very different insect in Africa. It is an assassin bug that lurks near termite nests. So long as it is covered by inedible shells of victims, the bug is in little danger of being recognized as food by a larger predator, and can stand in full view while waiting for the next termite to come along.

Virtually every country has assassin bugs of some kind, for about twenty-five hundred species successfully pursue this way of life. Many are black or dark brown. Most are agile and stealthy, too. The adherents of jump-and-grab see their victims from a distance. They turn to face the approaching target and adjust their position to keep it straight ahead. Often the predator makes quick dashes, then stands still again, until the prey is within leaping range and can be pounced upon.

Related ambush bugs of about one hundred kinds, mostly tropical, differ in having slightly knobbed, rather than thread-like, antennae and are generally yellow or some other pale color. They tend to be proponents of sit-and-wait, often resting on flowers so similar in hue as to be almost invisible. This

An assassin bug with a "cogwheel crest" on its back waits motionless for some unwary insect to come within reach, then pounces and drives its sharp beak into the victim.

deception lets the bug wait for a manageable victim to come along.

Both assassin and ambush bugs have front legs highly specialized as grasping tongs and a strong sucking beak. The grasp need not be large enough to reach around a victim, nor must crevices in the prey's body be found for the claws of the predator. These true bugs rely instead on special adhesive pads just above the foot region of each leg. The pad consists of as many as eighty thousand short hairs made sticky with a film of oil. The bug holds on with these hairs much more firmly than a fly can as it clings upside down to a ceiling. Using its strong grip, the bug probes for a thin place through which to push its beak and inject its lethal venom. The prey struggles, twitches, and goes limp in ten seconds or less.

Until 1899, many households in the Northern Hemisphere welcomed a jet-black assassin bug indoors. Even its wingless nymphal stages were tolerated, although they have a thick coat of lint and dust particles adhering to their backs. Known as the "masked bedbug hunter," the insect earned approval by ambushing bedbugs and feasting on other then-common household pests. But one night in that year, a woman in Washington, D.C., received a painful stab on the lip from an assassin bug of this kind. She had probably turned in her sleep and squeezed the bug that was policing her pillow. The newspapers, however, made much of the incident, calling the insect the "kissing bug," and soon even a mosquito bite anywhere near a human mouth was blamed on the black assassin.

A larger assassin bug of tropical South America remains a real menace to sleeping people. It goes out of its way to suck human blood, and spreads the trypanosome infection that causes Chagas's disease—a serious affliction that commonly begins with swelling of liver, spleen, facial tissues, and conjunctiva, then becomes a chronic anemia with heart involvement. Charles Darwin recorded being attacked during sleep in the village of Luxan, a short distance south of Lima, Peru. Yet his entry dated March 25, 1835, in the journal of the voyage of H.M.S. *Beagle,* shows his continuing curiosity about this "black bug of the Pampas":

It is most disgusting to feel soft wingless insects, about an inch long, crawling over one's body. Before sucking they are quite thin, but afterward they become round and bloated with blood, and in this state are easily crushed. One which I caught at Iquique, for they are found in Chile and Peru, was very empty. When placed on a table, and though surrounded by people, if a finger was presented, the bold insect would immediately protrude its sucker, make a charge, and if allowed, draw blood. No pain was caused by the wound. It was curious to watch its body during the act of sucking, as in less than ten minutes it changed from being as flat as a wafer to a globular form. This one feast, for which the benchuca was indebted to one of the officers, kept it fat during four months; but after the first fortnight, it was quite ready to have another suck.

The undiagnosed disease which so troubled Darwin in his later life may well have been contracted at this time.

Discoveries made in the twentieth century have made us aware of extra hazards from insect specialists if we get ambushed as they search for food. A sucking insect particularly, but a biting one too, can often transfer some agent of disease to which humankind is susceptible. Quite a few such agents have found ways to ride in or on insects since the six-legged creatures began to benefit from the mammals and birds that share their world. Generally, we dread the unseen microscopic parasites more than the visible insects that furnish the free transportation.

4

Survival of the Prey

TOGETHERNESS CAN BE WONDERFUL, AS LONG AS BOTH PARTNERS
gain from their proximity. But such balanced benefits are
scarce among the myriad interactions between one kind of
insect and another. Usually the members of one species lose
if they associate too closely with those of another. We identify
the loser as the prey of the predator, or the host of the parasite.
We laud as beneficial anything that gains by eating aphids or
caterpillars that feed on some plant we cherish. And we admire
the sensory refinements of the gainers, as they locate and then
take advantage of the losers.

But the gainers can lose too, if they are overly successful.
Their perpetuation depends upon the perpetual availability of
losers. The aphid and the caterpillar must continue to transform nourishment from plant form into animal tissues that the
predator or the parasitoid can use. Raymond L. Lindemann,
an American ecologist, looked into the logistics of this rela-

tionship in prairie populations, and in 1942 came up with a general rule. His conclusion has been tested many times since, and found reasonably true. He found that the carnivores together can gain only about 10 percent of the energy available in the herbivores they feed on, directly or indirectly, and will harm their own future if they destroy more than about a third of the reproductive capacity among the plant-eaters. Similarly, the herbivores can accumulate in their tissues only about 10 percent of the energy available to them in their food plants. A third of the vegetation may be wasted, but two-thirds should remain to perpetuate the supply.

"Enough" is the key word. The gainers must get enough, but enough of the losers must be left unharmed. Inherited patterns of behavior among predators and prey tend to stabilize these relationships. Commonly the victim is a young herbivore, rather than one that has survived to maturity: the young are more numerous, less wary, less well fitted to defend themselves, and have not yet taken as much energy from plants. Or the predators capture adults that have already reproduced as soon as they show signs of slowing down. This too conserves the green plants, in which photosynthesis captures the solar energy for the entire community of life.

All prey have ways to avoid being eaten that are at least twice as effective as the refinements that let the gainers take a toll. The potential prey may be expert at escaping, or effective in fighting off a predator. The possible victims may adjust their program of development to become available abruptly, in such numbers as to overwhelm the appetites of predators and leave them satiated, while the prey reproduces unmolested. Possession of a stinger or a poisonous body can help spectacularly. Sometimes success in survival comes merely by warning an attacker, relying on the fact that most predators have lived long enough to learn. The memory of a previous near-disaster may block the next move in the same direction. An inherited strategy that helps a prey species to survive in numbers is merely to stay perfectly motionless, silent, and odorless, or to blend as perfectly as possible with the surroundings, or to appear inedible.

No one has yet proved that a single insect knowingly plans its moves. Nor have many insects an opportunity to learn by watching their elders. Perhaps none remember. Yet in the long evolutionary interaction between predator and prey, the potential victims continually refine their means of surviving undetected and their ways of escaping when the predator gets too close.

ESCAPE!

One of the oldest ways to vanish is still magic after 350 million years. We see it performed every spring in New England, when we join friends who have tapped the sugar maple trees. Motes of black fleck the surface of the sap collected in the bucket. One wave of a hand and they are gone. The outdoor motes are snow fleas—springtail insects of ancient lineage. Each one routinely holds the tip of its tiny abdomen curled under and forward, tensed like the snap-loop of a mouse trap against a catch. At the slightest provocation, the springtail slips the catch. Its abdomen straightens out, flicking the whole body two or three inches into the air. The crowd of insects on the sap in the bucket becomes a meaningless speckling of separate motes elsewhere, no speck as much as a quarter of an inch long.

The simplest insects on earth take the second oldest way to flee from danger—bristletails simply run and hide. We see them on the rocky seacoast almost every time we picnic there. Others, inland, dodge us on boulders and tree trunks, where they find algae and lichens to eat. Bristletails also skitter to safety from mountain climbers in the Alps, at elevations as great as ten thousand feet. A few kinds, half an inch long, inhabit human homes, where they are known as silverfish, while others prefer bakeries, where they are called firebrats. Smaller kinds inhabit the nests of ants and termites. These perpetually wingless insects run on foot all over the world, as they have since late Devonian times.

A neighbor who had never encountered these escape artists before telephoned to ask our help. He had found two of them in the bathtub, unable to scale the slippery sides. "They travel like greased lightning," he insisted, "and the grease comes off on your fingers!" The "grease" is overlapping silvery scales (hence the name silverfish), which rub off easily and make the insect slippery. It is also distinguished by two long feelers at the front end and three at the rear.

We urged him to flush the two captives down the drain—and watch out for more of them in his library, since bristletails are partial to the glue that holds books together, as well as to leather goods and foodstuffs. They will gnaw the surface from glossy magazines. Usually the behavior of a bristletail gets it to food and away from danger with a minimum of adaptive features.

A cockroach is even better fitted for scurrying to safety. Its legs are longer in proportion than those of a bristletail, and swing strongly almost parallel to the ground. Long, mobile antennae and big compound eyes help the creature direct its course, and its flattened oval build lets it take refuge in narrow spaces. It can even feel its way backward by means of a posterior pair of short appendages. Yet the roach's best means of survival is in its avoidance of light in which it can be seen. By exploring chiefly at night, it is far less likely to encounter a predator that relies on vision.

We used to believe that cockroaches could not fly. Only once had we even seen one use its excellent wings. It was a native American roach, which had been resting in the narrow space atop a darkroom door. When we opened the door suddenly, the insect leaped into space, spread its wings to brake its fall, and closed them again over its back once it landed on its feet. The wings served as parachutes, and seemed to have no role in such routine activities as travel to new locations.

Giant cockroaches in the tropics may be exceptional, as we discovered one day when we accompanied an exploring scientist, Dr. Harold Trapido, into a cave in Panama. He wanted to collect pregnant vampire bats for an embryologist at Cornell University. Dr. Trapido led the way, commenting that the dry season offered the only chance to enter through a long tunnel to the major chamber where the vampires slept during daylight hours. Even then a steady stream of water eighteen inches deep coursed down the V-shaped bottom of the tunnel. During the rainy season, the water would rise six or seven feet, almost to the roof of the passage.

"You'll soon have a narrow dry ledge to walk on," Trapido

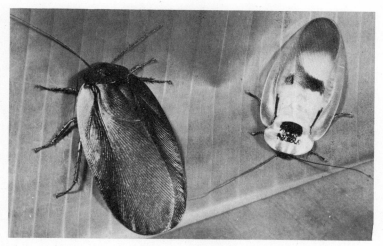

Tropical cockroaches come in giant sizes and often shelter for the day in the shade below a banana leaf. These specimens were photographed in Panama.

assured us. "It starts fifty yards or so from the entrance." So one Milne (Lorus J.) took the other pick-a-back. One pair of jungle boots stayed dry, while the other, larger pair sloshed through the stream. Each of us carried a spare flashlight in a side pack, and a powerful electric lamp on a headband connected to batteries on a belt.

For a few yards, the trip seemed as uneventful as though we headed into a commercialized cavern in the United States. Then we noticed a few immature cockroaches of a kind that measures four to five inches in length when full grown. The juveniles clung to the cave walls, wildly waving their antennae, and shifting position slightly whenever a headlamp beam centered on them. "There's an adult!" Margery cried, freeing one arm from Lorus's neck to point. Dozens more now were discernible farther down the passageway, as we shone our lights ahead.

Suddenly the roaches reacted to something, and started to run for new positions on the walls and roof. Many adults dropped or leaped into flight, winging straight for our headlamps. Trapido, at least, had two free hands to beat them off. Margery had a free arm to deflect the insects that struck her. But Lorus heroically had to tolerate a succession of six-inch roaches clambering over his brows, eyes, nose, and mouth, as he held his wife above the water!

Once the cavern was reached, we saw no more roaches, and no bat, disturbed by the lights and our presence, flew close. We were now convinced that cockroach wings will sustain directed flight. Yet this behavior is probably a last resort at night, because the bats then are actively echolocating their insect prey on the wing. Running would be far safer for a roach in darkness—a better way to keep from being eaten.

The flea, the flea-beetle, the grasshopper, and the cricket all move more abruptly in a familiar way when they are disturbed. Disproportionate enlargement of muscles in the hindmost pair of legs gives them the power for a quick leap.

Sir D'Arcy W. Thompson finds more to marvel at in the human high jump, because the jumper's center of gravity goes under the measuring bar, while the body, "bit by bit, *goes over* it." Insects perform no such ingenious tricks. In consequence: "... the grasshopper seems to be as well planned for jumping as the flea, and the actual heights to which they jump are much of a muchness; but the flea's jump is about 200 times its own height, the grasshopper's at most 20–30 times; and neither flea nor grasshopper is a better but rather a worse jumper than a horse or a man." We should not overlook the fact that the leap of a flea is as often toward a meal as away from danger. The grasshopper, the cricket, and the flea-beetle jump to save their lives.

An insect earns less honor from a respectable leap than from an active defense. Lorus J. Milne learned this at a tender age by picking up a short-winged grasshopper in Florida. The huge insect, called a lubber because it moves so languidly, seemed reluctant to jump, as well as unable to fly. It scarcely struggled. Slowly, however, the insect changed the position of its hindmost pair of legs until the knee joints were well above the human fingers. Suddenly the insect drove its shanks downward. Spines on those shanks, as sharp as the teeth of a saw, opened human flesh into two bloody, painful gashes. A less determined insect-catcher might have let that grasshopper go free. The child chose, however, to use his other hand to seize the trophy again in such a way as to immobilize those strong hind legs, and saved the insect for a place in a museum tray.

Many an insect, small or large, is adept at spreading its wings and flying from danger—or just to travel. By day they leave one hazard for others at least as deadly, such as a dragonfly or a bird. By night, insectivorous bats may be a constant menace. Some of the night-flying moths are equipped to cope, for they possess a pair of special sound detectors called tympanic organs. These may respond over the whole spectrum of sound from eight to beyond one hundred thousand vibrations per second. The tympanic organs send urgent messages to the moth's nervous system whenever the pulsing calls of an echolocating bat stimulate these special hearing organs. The moth makes a power dive to the ground, and stays there quietly until the bat flies past. Although very small moths apparently lack tympanic organs, many species that a bat would relish make use of this system routinely.

The moths that mature from a particularly noxious cutworm found in gardens are among the most successful in avoiding bats. They will demonstrate this saving behavior if tested in a cage under the flight path of bats emerging at nightfall from a roost in an old barn. Scientists have confirmed the internal mechanism by dissecting out the tympanic organ from a cutworm moth and connecting it to electronic equipment. Any messages along the nerve from the organ become sounds audible to the ears of the human experimenters. Tested in this way, the bat detector of the moth begins its warning while the bat is still a hundred feet away and thirty to fifty feet higher up; it continues to send characteristic messages until the bat passes thirty to forty feet beyond the observers, still at the same altitude.

Flight by adult insects with good wings, like running and leaping by insects of any active stage of development, reminds us of comparable behavior in animals with a backbone. Yet none of the insects shows the evasive action that is so familiar when a frog feels threatened while resting on land beside a pond. The amphibian leaps for the water, often with a cry of alarm, and dives to safety. Only a whirligig beetle, already swimming half immersed in the water film, dives to avoid being caught. Yet water boatmen (corixids), which are true bugs of much smaller size, as adults escape pursuit in the opposite direction. Each of these common insects feeds delicately submerged, on a mixed diet of minute water plants and animals, sucked in through a comparatively feeble beak. Such modest meals are available everywhere—in brackish pools along the sea coast, in bird baths, in quiet parts between cascades in streams, and in mountain pools as high as sixteen thousand feet. In none of these situations is a boatman particularly adept at escaping a backswimmer. Boatmen, in fact, provide either routine fare or emergency rations for insect predators and fishes of the freshwater community. Adult boatmen keep a bubble of air under their overlapping wings, which gives them considerable buoyancy; it forces them to hold on with their legs while feeding on the bottom. But if really threatened, the insect lets go. It rises to the surface so rapidly that it pops through the water film into air, and generally flies away. Sometimes great swarms of these little bugs burst forth and travel from one pond to another.

Water striders are often a source of nourishment for fish, but even as the fish tries to gulp down this nourishment, the strider is likely to leap away. Its middle and hind pairs of legs then work simultaneously. In consecutive hops, they toss the strider as much as four inches into the air, generally in a forward direction. Rarely, however, does danger of this kind induce a strider with wings to fly away. (Many species, as adults, have both winged and wingless members.) The fliers are well fitted for dispersal, while the nonfliers are strictly stay-at-homes. Curiously, the winged striders fly mostly at night, sometimes for miles, generally in autumn or when their watery world threatens to dry up or choke with vegetation. Occasionally a strider turns up in a birdbath and stays until the first bird arrives to get a drink. More remarkably, the far travelers among these insects of the water film are mostly males. After surviving a winter in some new location, they use their own wing muscles as a source of nourishment, prolonging their lives and using the energy while hunting for food and mates.

FIGHT!

Collectors of insects distinguish between members of orders with biting mouthparts and those that have adaptations for sucking, usually after piercing a plant or some prey. Biting

jaws obviously should be avoided, particularly if the head of
the insect is well developed and provides space for powerful
muscles to close those jaws. Sharp-tipped mandibles may perforate human skin, and even blunt ones can provide a painful
pinch. A "pinch bug" generally proves to be a stag beetle,
belonging to a family represented on every continent that has
forests. The roots of trees provide the nourishment needed by
the C-shaped grubs of stag beetles. The adult insects include
many that appear almost as ready to stand and bite in self-
defense as to run, rather awkwardly, away. The males of many
species appear particularly formidable, with strong jaws that
seem definitely oversize.

Sir Julian S. Huxley of King's College, London, found stag
beetles especially intriguing because, while size varies considerably among both sexes of a species, the prominent jaws of
a big male are disproportionately larger than those of a small
male. Since the adults have stopped growing, these differences
in development occur during the feeding, larval stages. A well-
nourished larval male becomes a giant with huge jaws, while
a male larva that ceases feeding and pupates at small size
becomes a lesser adult, with far smaller jaws. For example, the
jaws of the common stag beetle of Britain will be almost half
as long as he is if his head plus body measures two inches, but

A comparison series of the European stag beetle shows a female at the right and
three sizes of males on the left. The males that, as larvae, received more to eat
and grew bigger before pupating, show progressively larger jaws and head.

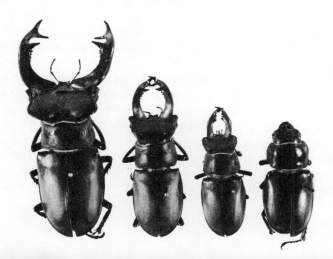

only a fourth as long as he is if he measures only 1.3 inches. One wonders what sort of monstrosity such an insect might become if he were three inches long, perhaps with jaws longer than his body. The answer is shown by any male that is 1.50 inches or longer from the back of his head to the tip of his closed wing covers—the parts that bear the locomotory legs and wings and that contain the digestive and reproductive organs. The length of his jaws plus his head is always exactly equal in length to these more posterior parts of his body. The biggest males are not so penalized as has been believed, for they possess larger heads containing jaw muscles to take care of the bigger jaws.

Nothing in the behavior of stag beetles seems to reward males for bearing armament of such disproportionate dimensions. As we watched our North American stag beetles in action, we suddenly realized that two unlike features overlapped in earlier measurements. The biting mechanism, which could be included in the overall length of the insect, has its muscles in the head. Jaws and head should be a unit. The thorax, with its locomotory legs and wings, and the abdomen, with digestive and reproductive organs well protected below the closed forewings, should be a second unit. These two regions show complete consistency for the females of the European stag beetle. Her jaws plus head are regularly 23 percent of the combined length of her thorax and more posterior parts— regardless of whether the combination measures 1.00 or 1.34 inches. Males have head plus jaws equal to 40 percent of the total for thorax and closed forewings, if this total is only one inch long. The percentage rises regularly until the male attains 1.50 inches in this dimension. Then, or at still larger size, his jaws and head attain 100 percent exactly of the posterior portion of his body.

Being nipped in the dark by some small six-legged creature can provide sudden understanding. One example in our experience occurred during a beach picnic on one of the Elizabeth Islands off the southwestern tip of Cape Cod, Massachusetts. While other biologists from the marine laboratory huddled close to a bonfire and sang to warm their spirits, we stumbled through the night to where we had left sweaters while the evening was still mild. Sliding hands into sleeves, Lorus encountered something that ran away and nipped him as it went. The nips came from the retreating end of the insect. Instantly he knew that it was an earwig, without having met it or any of its order previously. In this case, the nipper was the seaside earwig, a wingless insect with a somewhat flattened body about one-and-a-half inches in overall length. The tip of its abdomen bears a pair of pincers. Tradition holds that the name "earwig" comes from a resemblance between these terminal append-ages and the instruments once used to pierce women's ears for earrings. More plausible, perhaps, is the claim that earwigs will take shelter in the ears of people sleeping on the ground. The least likely explanation is that someone noticed the second pair of wings on earwigs that fly as adults, and recognized a similarity in form between the expanded wing and the human ear.

Any earwig will use its abdominal pincers to fight a rear-guard action while it slithers away. The seaside earwig some-times uses these nippers like an extra pair of jaws—much larger than the paired mandibles at the mouth—to grasp a beach flea or some other small creature. The earwig then twists its flexible body and brings the hapless victim into posi-tion to be eaten.

Earwigs with large, fan-shaped hind wings use their abdomi-nal forceps to help themselves fold up these wings and fit them under the scalelike fore pair, which double as wing covers. The smaller European earwig, which has been introduced and spread in New England, will demonstrate this behavior if watched closely at night, when it occasionally flies about. Those that invade our home, and run for safety when discov-ered, rarely get a chance to show what their rear pincers can do.

STING!

We admit to getting stung occasionally, sometimes because we misjudge the response to be expected from an insect. The greater role of scent in inducing defensive action was im-pressed upon us one afternoon when we found a nest of white-faced hornets in an unusual location. Never before had we seen one of these globular edifices, a foot in diameter, in a shrub so close to the ground. We set up the camera on its tripod, zoomed-in the lens to catch the detail in the doorway area, and waited impatiently for several wasps to crawl about there at once, as single individuals did occasionally. Tired of crouching over, Lorus blew one sustained breath toward the nest a dozen feet away. Out flew a hornet straight for the photographer and stung his forehead where his mouth had been a moment earlier, before his eye returned to the camera. One sting, one lesson, and off the wasp flew on other business for the colony.

But sometimes, one stinger or even a hundred of them is not enough. A week later, an early frost immobilized the armed defenders of the paper nest. A skunk made a nocturnal meal out of every inhabitant, young and mature. Now we know why most white-faced hornets inherit the habit of building beyond the reach of pedestrian animals.

Many ants defend themselves vigorously with a venomous stinger. Moreover, they place their poison with extra care by using their jaws to hold on to any intruder while jabbing it with the stinger at the opposite end of the body. Few people will stand still and make a calm appraisal of this technique if they chance to stand close to a nest of fire ants, now widespread in the American Southeast. A fire ant uses its jaws to pinch up a fold of human flesh, while stinging it repeatedly. The venom hurts like a burning flame, and, generally, a dozen or two fire ants are making their points simultaneously on the same ankle or leg, if not on both. Southern comfort would have lasted longer if the fire ants had not been accidentally introduced from South America, where they probably had more to defend themselves against.

The same technique of bite and sting is particularly well developed among the bulldog ants of Australia. These vigorous insects, often exceeding an inch in length, will leap calf-high from hard ground to warn away any person or animal approaching their nest. A single experience convinced us that their venom is more potent than that of any fire ant.

Only once have we feared that human bulk alone might not suffice to dilute the effect of an insect sting. The stinger was a huge ant in Panama, which we caught in the dense shade of the rain forest, and flipped into a cloth bag for transport into better light. When confronted a few minutes later by a newly fallen tree across the trail, we forgot the captive ant, and carefully spread the cloth bag over the rough horizontal trunk to have a safe surface for clambering over the obstacle. The ant stung the hand that nudged her through the cloth. The sensation had the sharpness of a wasp sting, neither more nor less. But in a minute the whole hand became numb. The loss of all sensation spread past the wrist. Soon the anesthetic action reached the elbow. In less than fifteen minutes, it extended almost to the shoulder. Should we apply a tourniquet? Where do you get advice when the only primates within a mile are capuchin monkeys and howlers? Fortunately, the numbness faded in the reverse direction at an equal pace and, in less than a day, the hand felt none the worse.

The bumblebee near home that stung us through the cloth net when we carelessly pressed against her offered only a prickly reminder that she was there. Later, we freed her unharmed. But any mouse that came to rob her nest might receive a different message as her charge of venom spread through its smaller body.

In more ways than one, we feel pained when we so disturb an insect that it responds by stinging. Usually their tolerance of human presence convinces us that unprovoked attack is most unlikely. Thousands of honeybees can be robbed of their laboriously gathered sweet stores without a single individual reacting violently. No worker bee need get so excited that she commits suicide by driving her stinger into human skin and partially eviscerating herself to leave the poison gland contracting alongside her implanted dart.

Stingers and potent venom do not always constitute a sufficient deterrent. Robber flies of several kinds often loiter near beehives, where they can pounce on bees, hold them firmly in long hairy legs, and somehow settle on a leaf without losing hold or getting stung. Thrusting its sucking mouthparts into the body of the captive bee, the robber fly sucks out the nutritious contents and discards the empty shell. Sometimes a pile of empty bees accumulates below a favorite perch—much to the distress of the beekeeper. Equally distressing in Europe and Africa, which both had wild honeybees long before humankind arrived to domesticate these insects, are the bee-eater birds that swoop like swallows in pursuit of bees and wasps. Bee eaters must possess the same kind of tolerance as the American toad, which seems capable of ingesting any stinging insect that comes within tongue range, without side effects.

POISON!

A chemical weapon can dissuade insect-eaters from attacking many insects that have neither a stinger nor strong jaws. A captured grasshopper will "spit tobacco juice," by exuding the brown and often malodorous contents of its crop. A shield-shaped stink bug will leave its nauseating odor on the raspberry from which it has been frightened. The large water beetle *Dytiscus* will eject fecal pellets like projectiles toward an approaching fish, no doubt fouling the water with a repellent wake.

The most refined version of chemical defense belongs to small ground beetles that are widespread in the Americas from Canada to Argentina, and somewhat larger ones in Africa, Asia, and the East Indies. Known appropriately as bombardier beetles, they react to disturbance by firing off explosive charges from the rear. Every discharge makes an audible "pop" and produces a little cloud of brownish gas. Occasionally the clouds combine while the beetle hurries on. The half-inch insect, powder-blue on wing covers and orange elsewhere, becomes memorable to gardeners who shock a bombardier or two with daylight by overturning a board or stone under which the beetle has been hunting.

Once in a great while we meet a bombardier with a six-shooter. More often, the insect's gun is emptied by four or five little explosions in quick succession, although one beetle, examined by Thomas Eisner of Cornell University, responded to continuous stimulation with twenty-nine shots before giving up. Far fewer are needed to convince anyone that the gas feels warm on a fingertip, and that the skin turns brown as though it had been swabbed with a strong iodine solution. Large bombardiers are credited with burning a human hand so severely that only a few specimens can be captured with thumb and forefinger on any single day of collecting. All bombardiers earn high scores for marksmanship, making direct hits in almost any direction by turning the flexible nozzle of the defensive organ.

The chemical and anatomical basis of the defense system of these spectacular beetles was clarified for the first time in 1969. A cluster of gland cells transfers secretions through a duct into the first large chamber of a two-compartmented organ. The solution there consists of 25 percent hydrogen peroxide and 10 percent hydroquinones, which are the active ingredients. When the insect is alarmed, muscles surrounding the reservoir squeeze some of the solution into the second chamber, a sort of vestibule that leads toward the nozzle. The vestibule walls secrete enzymes that facilitate an instantaneous explosion: peroxidase causes the hydrogen peroxide to decompose into water and free oxygen, while catalase helps the hydroquinones change into toxic quinones and hydrogen. At the instant of the explosion, the hydrogen and oxygen combine to form water and release energy. The temperature of the discharge rises to the boiling point of water, with enough heat left over to vaporize almost a fifth of the discharge. Out goes an extremely hot jet of steam and minute droplets of quinone solution. The research team at Cornell University, in providing these details, concluded that "a bombardier beetle can make itself felt thermally, even where the chemical 'message' cannot get through." Whether the bombardier is insulated internally against being burned by heat from its own gun remains to be discovered.

The active compounds that a bombardier uses in its gun have found favor among humankind only recently. Hydroquinone became an ingredient in photographic developers about a century ago, because it could help transform a latent image in exposed emulsion into a visible pattern composed of silver grains—a picture, as though by magic. Photographers who did their own processing had to learn to keep their fingers out of the developer, or have them stained dark brown with toxic quinones. Hydrogen peroxide now provides both the propulsive power and the heat for a device patented as recently as 1967, which delivers hot lather for shaving.

Quite unrelated beetles discourage most insect-eaters from attacking them by secreting liquids rich in cantharidin

($C_{10}H_{12}O_4$), a substance that causes an immediate burning
sensation on human skin, followed by watery blisters. It is also
a potent stomach poison if it enters the body in other ways.
Some oil beetles (of family Meloidae) exude the potent liquid
from their knees when disturbed in any way; these are "blister
beetles." Most of their two thousand relatives store the poison
in tiny chambers within their soft wing covers and elsewhere,
ready to affect any predator that squeezes the insect before
swallowing it.

Physicians of the Middle Ages discovered that a drug could
be prepared from the dried wing covers of a common oil
beetle from southern Europe, if these were pulverized and
applied in a lotion to the body surface of the patient. The
mixture would redden the skin as a "rubifacient," or produce
a wonderfully distracting patch of blisters as a "vesiculant."
Soon, blistering ranked with bloodletting as a standard treatment for illness of almost any kind. The practice continued
until late in the nineteenth century. Many modern apothecar-
ies, in fact, still stock a jar of broken beetle wing covers, because veterinarians occasionally call for this ancient remedy in
treating horses with damaged legs. Known by the common
name of "Spanish fly," these dried remains of the oil beetle
Lytta vesicatoria have not yet been replaced by old-style mustard plasters or new-style electric hot pads, infrared lamps, or
ultrasound. The belief that adding a small quantity of the drug
to the food or drink of a prim lady, which causes irritation in
the urinogenital tract, will act as a guaranteed aphrodisiac has
been decried, since large doses produce lasting damage and
can be fatal.

The insects that secrete cantharidin include not only oil
beetles, but also soft-winged flower beetles (of family Mala-
chiidae) and some weevils. Oil beetles and soft-winged flower
beetles commonly associate in considerable numbers, completely exposed atop clustered flowers from which they gather
pollen and nectar. Like certain species of venomous tropical
fish, many of these beetles have a striking appearance. They
come in conspicuous colors of metallic green and blue, often
with bold patterns of bright orange spots on their backs.

The same daisy may simultaneously attract an oil beetle, a
soldier beetle boldly marked in black on red or orange, and a

glittering leaf beetle. All three tend to ignore the approach of a predator or a person, and each has its own unique poisons to dissuade an attacker. The soldier beetle has white secretions ready for discharge from glands along its sides, and these can be "milked" by gentle procedures to obtain samples of the solution. Its effective ingredient seems to be unique. The special molecule is based upon an acetylenic acid: such substances have been discovered in certain fungi and flowering plants, especially members of the daisy family (Compositae). Does the soldier beetle or its carnivorous larva seek out a plant source of acetylenic acid, or can it synthesize its own as needed? So far, only the beetles have the answer.

Many of the leaf-eating beetles of family Chrysomelidae possess brilliant colors as adults, and show themselves ready when disturbed to exude copious amounts of poisonous fluid. Often the effective ingredient is a cardiac glycoside, a "heart poison" similar to or identical with the substance extracted for medical use from the purple foxglove *(Digitalis purpurea)* of western Europe. The poison is present in the adult, the eggs, the larvae, and the pupae. Recent experiments prove that these beetles, at least, can synthesize the ingredients for their chemical defense despite feeding for successive generations on plants that lack all traces of heart poisons. The quantities present in the eggs are far too small to account for the larger amounts in the later stages of development.

The monarch butterflies, which are so familiar over most of North America, do get their poisons second-hand. The pregnant females hunt out milkweed plants on which to lay their eggs. Monarch caterpillars, striped narrowly around by bands of yellow, black, and white, tolerate the toxic glycosides so many milkweeds possess in their milky juice. A caterpillar will temporarily abandon its munching of milkweed foliage to drink the juice itself, if this is offered inconspicuously in a spoon. So much of the poison remains in the body of the caterpillar that if it is eaten by a bird, the meal causes either a life-saving regurgitation, or death by cardiac arrest. A bird, a mammal, or a reptile that has had to vomit after eating a hairless caterpillar with these markings generally remembers not to eat another.

With virtually no need to avoid predators, a milkweed caterpillar can concentrate on feeding. It grows rapidly and sheds its skin a final time to reveal a handsome chrysalis that is pale green with gold and black dots—and equally poisonous. The chrysalis turns color, revealing through a glassily transparent covering the orange and black markings of the wings on the developing butterfly. The insect emerges in due course, dries its expanded wings, and flies about. Its body is still booby-trapped for predators. An experienced bird that seizes the monarch by the wings is reminded by the bitter flavor to release the prey. But an inexperienced bird, or one that is extremely hungry, may ignore this warning, carry the monarch to some perch, pluck off the bitter wings and drop them, devour the thorax with its waving legs and wing stumps—perhaps the head as well—then gulp down the abdomen. It is in the abdomen that the poisons are most concentrated. Vomit or die then becomes the chemical imperative.

WARN!

Almost a century ago, while observing insects in Brazil, the German naturalist Fritz Müller first recognized that members of unlike species with a similar defense system tended to resemble one another. He concluded that bees and wasps benefit by having comparatively uniform stripings of black and yellow, since insect-eaters learn quickly to avoid them. Otherwise a hungry bird or reptile would have to recall instantly a whole array of different patterns, each one learned by painful experience, to avoid the insect before getting stung again. Müller insisted in 1879 that in this way "each step toward resemblance is preserved by natural selection."

Many of the beetles that are admired as "living jewels" prove to be highly poisonous. Their larvae too may contain the same or equally deadly substances, without bearing the bright colors and distinctive patterns that might warn a predator with good eyes and memory that this is a tidbit to ignore.

A conspicuous pattern of colors, an audible buzzing, or any other obvious feature shown by a creature that has armament can help to ensure that it will be recognized and avoided on subsequent occasions. Fewer, in fact, will be maimed or killed in teaching the first lesson, if insects with similar weapons look and act much alike—and move about freely, letting the warning be noted.

We might suggest that squeaky beetles mimic one another. Each one could reinforce a conditioned reaction in a bird or bat, if the insect produced both a squeak and a simultaneous exudate of disagreeable material. Many a burying beetle squeaks if captured, and quickly discharges a liquid from its anal region.

Many (perhaps most) of the tiger moths, assigned to the unrelated family Arctiidae, begin emitting sounds of their own when they hear an approaching bat. Clicks, rapidly repeated, arise due to the buckling action of a special thin area on each side of the moth's thorax, caused by muscle contractions inside this part of the body. Compared to other insect sounds, these clicks seem faint. At half an inch away, they are no louder than a food blender working at slow speed. But their pitch is predominantly in the ultrasonic range above twenty thousand cycles per second. That the bats hear and react to the moth sounds is evident from the regularity with which the flying mammals veer away from the clicking moth. Probably experience teaches each bat quite early in its life that a clicking moth is distasteful, if not poisonous. The warning quality of the moth signals could also be inferred from the records kept by biologist James H. Fullard of Carleton University in Ottawa, Ontario. Near the town of Brockville, he found that tiger moths of species that are active in spring remain silent even when stimulated by a recording of an insectivorous bat's hunting call. No bats are active so early there. But between mid-June and early autumn, in bat season, tiger moths in the same locality click regularly when stimulated in the same way.

Two closely similar tiger moths in the same genus display this difference. One *(Phragmatobia assimilans)* "is a spring moth emerging early in May and disappearing before June," according to Fullard. It is silent when stimulated, although it possesses in the side walls of its thorax a generous vestige of the sound-producing mechanism (tymbal). The second species *(P. rubicosa)* emerges in late July and remains active through August. It emits about 650 pulses per second of high-frequency clicks from well-developed tymbals. Fullard suspects that the silent species had sound-producing ancestors, but that they changed their emergence time to spring for some reason and avoided the period of bat activity. With no further need for functional tymbals, they gradually lost the use of this special organ and then much of the organ itself.

HOAX!

Most predators learn to avoid a bee or a wasp striped with black and yellow—or anything resembling an ant—which could bite and sting, or certain color patterns of butterflies if these generally turn out to be distasteful or poisonous. These warnings are so effective that insects with no defense system of their own may escape attack if they resemble protected species in shape, coloration, and behavior. The English naturalist Henry Walter Bates noticed many examples of this kind among the butterflies of the Amazon basin. For him, the resemblance was significant only if the harmless mimic moved about when the protected look-alikes ("models") dominated the scene, able to teach any predator a lesson. With the odds in favor of getting a distasteful or dangerous mouthful, the insect-eaters would ignore the model and the edible mimic equally.

It seems strange that until 1861, when Bates drew attention to mimicry of this type in South America, no one had noticed the phenomenon elsewhere. Yet Eurasia and North America have long been homes of the harmless drone fly. And any sharp-eyed youngster can learn to distinguish the fly from the honeybee it so closely resembles in size and coloration. The larger head and short bristly antennae of the fly differ from the smaller head and elbowed antennae of the bee much more obviously than the fly's broader waist and single pair of wings contrast with the constricted waist and two pairs of wings on the bee. You can snatch a drone fly from a flower with a quick move and hold the insect unharmed in a closed fist. The fly buzzes much like a captured bee, but lacks a stinger with which to follow the warning sound. If you open your palm, the fly may walk about for a few seconds, prodding the skin with its abdominal tip as though to sting more before flying away. (People who are unacquainted with a drone fly marvel at the bravery of a person who tolerates the insect without flinching.)

Yet there is a disadvantage to the fly's disguise: it is often caught and eaten by the same kinds of robber flies that hunt honeybees. We wonder if the flavor is indistinguishable. The fly maggot gets no rich diet of honey and pollen. Instead it develops in fetid sites of decomposition, such as the fecal material in a privy, extending its telescopic "rat tail" above any liquid, as a snorkel reaching into air.

The monarch butterfly has a mimic over much of its range in North America, although not in the Northwest or most of California and Nevada. The edible species is the viceroy, a slightly smaller insect, which gains a good deal of immunity

The scarce swallowtail of southern Europe is paler than the common tiger swallowtail butterfly, but similar in markings and with only a single color pattern. This specimen was photographed at Herculaneum in Italy.

from predators by resembling the monarch in color and pattern. Insect-eaters appear not to notice the black curving mark on the rear wings of the viceroy, a feature missing on a monarch, and thereby overlook a bit of nourishment that has developed at the expense of foliage on willow and poplar trees.

Birds that hunt butterflies as food seem equally deceived by look-alikes in the southeastern states. Wherever the distasteful pipevine (or spicebush) swallowtail is common at puddles along woodland roads, the insectivorous predators learn to avoid it. They ignore also the dark females of the widespread tiger swallowtail—blackish-brown with blue and greenish-yellow spots—because they resemble pipevine adults. Both dark and pale forms of the tiger swallowtail coexist in much of the Southeast, but male tigers show a definite preference for light females colored like themselves, canary yellow streaked with black. They ignore many of the dark-form females. Those few dark females that are impregnated gain so much protection by mimicking the distasteful pipevine butterfly that virtually all their eggs get laid. The much-mated pale females get caught and eaten before they go far. John M. Burns of Wesleyan University credits this difference with preserving the inheritance of dark-form tiger females.

HIDE!

Small relatives of the cicadas have their own way to avoid being eaten while they perch on grass or herb or tree, sucking out the nourishing juices from the plant. These insects grow barely more than a quarter inch long, but do so hidden amid a mass of froth. Known as spittle insects from this protective covering, or as froghoppers because of the shape of the body and ability to leap if uncovered, they effectively disappear without altering their other habits. Each insect exudes daily during its immature development a clear fluid from its anus, combining the material with the secretions from glands in the seventh and eighth segments of the soft abdomen. It blows exhaled air through the mixture from a special chamber into which the breathing pores open. The secretions stabilize the froth, which accumulates as a protection from sun, dry air, and

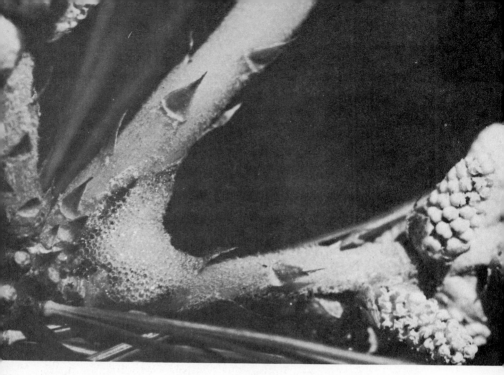

Spittle insects (or froghoppers) conceal themselves within a mass of froth, this one among needles and young cones on a pine tree. Few birds dip into the froth to eat the soft-bodied, immature bug that is sucking sap and producing the froth.

predators. Few birds will thrust a beak into the spume to capture its producer. That some investigate may explain why the froth is often known as "cuckoo spit." Some solitary wasps, however, show no such reluctance. They dive into the froth and drag out the spittle insects to use as food for young of their own kind. Left alone, a spittle insect gradually transforms into a small oval adult, holding its wings tentlike over its body much as a cicada does. The adult lacks the ability to produce the enveloping spittle and spends most of its days locating a mate or placing eggs on appropriate vegetation.

Natural hiding places abound, and serve insects of many

kinds—particularly those that feed on fungi or one another. A wonderful community of beetles finds refuge from predators, sunlight, and dry air under the loosening bark of dying or dead trees. Ground beetles may hunt there if the space is adequate, or if the insect's body is especially thin. One kind two inches long and an inch in width is known all over the world as the "violin beetle" because of its flattened body and narrow, forward-projecting parts; it is native only to Sumatra, where it seldom shows itself in the open, as though aware that some larger predator might eat it. Wingless camel crickets scavenge under bark, or reproduce in dark caverns underground as "cave crickets." Their antennae are extraordinarily long, their legs bristly with sensory projections, alerting the insect to dangers in the dark while a moment remains to leap elsewhere.

Bagworms and caddisworms build their own shelters and hide inside from danger. Bagworms are the caterpillars of moths belonging to a worldwide family. Caddisworms are the caterpillarlike aquatic larvae of caddisflies, a whole order of insects that is well represented on every temperate continent. The larval stages of both types of insects produce a sturdy tube, open or closed at the rear end, but quite obviously composed of materials that no insect-eater would normally devour. They use a saliva that is as waterproof and adhesive as rubber cement to attach new fragments to the shelter, or to spin out threads as uniform as those with which most moth caterpillars construct their cocoons.

The case constructed by a caddisworm does double duty. It hides the soft body of the insect from hungry predators, and, in places where the water current is too slow or unreliable to bring a continuous supply of oxygen, it enables the caddisworm to satisfy its respiratory needs by inconspicuously undulating its body within the case, propelling water in at one end and out at the other. This behavior can be easily observed if a caddisworm is evicted from its case and offered a short length of clear glass tubing as an acceptable substitute.

An evicted caddisworm will attempt to gain a new home by piracy, if it finds an opportunity. Repulsed from the front doorway of an inhabited case of suitable dimensions, the naked larva quickly tries to enter from the opposite end, biting the hindmost parts of the present occupant until the takeover is completed. Of course, the newly evicted larva circuits and repeats the move. Around and around the two go, until one shelterless larva gives up or is eaten by a fish, a diving beetle, or some other insect-hunter of the watery realm.

Construction of a new case takes longer. It requires special patterns of behavior that differ according to the family or the genus of the caddisworm. Some larvae commence by swinging the head while spinning a silk loop horizontally upon the silty bottom of the pond. When the loop hardens, the caddisworm steps on the nearest part and raises the loop into a vertical position around its neck region. One pair of legs holds the loop vertical while the larva cements pieces of plant or mineral material to the forward edge of the loop. Gradually a collar accumulates and lengthens, until the body of the caddisworm is hidden completely inside. Special leglike hooks at the posterior end of the larva hold the completed case while the caddisworm reaches out the front doorway to haul itself along or to grab food that it pulls back in to eat.

The tiniest of caddisworms—the "micros" of order Trichoptera—construct their hideaways entirely of silk and glue. They temporarily attach their small flat bags, each with two openings, to the lower sides of rounded rocks in riffles, where water flows through slowly. So small a caddisworm has no difficulty turning end-for-end within its flexible case and feeding alternately from the two openings. When the supplies of edible algae coating adjacent areas of rock have been grazed fairly clean from both doorways, the caddisworm frees one end of its case. The current swings the shelter, and the larva fastens it in this new position. If the grazing is still poor, the opposite end of the little bag is cut loose, and the case swings again.

A chewing insect, or one that preys on smaller creatures, may gain concealment while moving about if it carries a portable case that hides its body. If this covering resembles nothing that an insectivorous animal regards as edible, the case-bearer may escape detection altogether. The larvae of certain leaf beetles take full advantage of such deception by cementing together their own fecal pellets as building materials and fashioning thimble-shaped cases for their soft bodies. Until one moves on a leaf, it can easily be mistaken for the dropping of a large caterpillar.

CAMOUFLAGE!

In the long evolutionary interaction between prey and predator, the potential victims continually refine their means of
surviving and their ways of escaping when a predator comes
close. One of the most successful techniques takes full advantage of any chance resemblance between the edible insect and

*Many a treehopper resembles a thorn so successfully that no bird sees it—or
notices that the imitation thorn is pointing in the wrong direction for a product
of the plant.*

portions of the surrounding scene that hold no interest for an insect-eater. A treehopper bug may suggest only another thorn on a prickly branch. A stick insect, although kin to praying mantis and cockroach, seems to simulate a tangle of twigs. The famous leaf butterfly of Southeast Asia resembles a dead

A stick insect stands still or moves scarcely faster than a wind would sway it, and thereby escapes detection by day, feeding in darkness on foliage of trees and shrubs.

leaf attached to a plant. Some caterpillars have the outline and coloration encountered in a bird dropping. Many a moth remains imperceptible while it clings by day to tree bark, with the same general hue and irregularity of markings.

Such deceit by an insect relates to the fact that the hunters by day look for movement. A toad actually has a built-in "bug detector" among the nerve cells of its optic pathway. It blanks out the image of anything that stays still, to concentrate on any edible item that might move in the field of view. Except at night, "preadaptive resemblance" has some serious disadvantages. The prey species must remain perfectly motionless, silent, and odorless. This generally requires an insect with jaws to stop eating. But fasting wastes time in a short lifespan. So does cessation of activities essential for mate finding and other behavior essential to the burgeoning of the species. The herbivore continues life, but produces fewer offspring.

OVERWHELM!

The Methuselahs of the insect world are also the most noisy. They are adult cicadas, mistakenly called locusts over much of North America. Anyone who has ever caught a male cicada knows that the insect will buzz vigorously while held in the fingers. We occasionally hear this same "protest squawk" from an insect high in a tree, and notice that it ends abruptly. It tells us that a cicada-killer wasp has pressed an attack and quickly tranquilized its prey with a well-placed jab of the venomous stinger. But this sudden end to the cicada's socializing is infrequent and most of the time, the songs go on and on.

Cicadas that take either thirteen or seventeen years to reach maturity are the real experts at overwhelming the insect-eaters by sheer numbers. Every year there may be a brood of them somewhere in the country. But the area where each brood emerges—as many as 1.5 million to the acre—is separate from that of every other, and each area produces a new population of adults only in the one year of its cycle. The cicadas provide no annual resource upon which cicada-killers might build up a greater population. When a brood does mature, relatively few individuals lose their lives to predators before these eaters of insects want no more. The rest of the cicadas move and call

A sudden stop to the song of the cicada often indicates that a cicada-killer wasp has found and stung a victim. The wasp also uses the stinger to help carry the inert cicada, grasping it with its legs as well, while flying to a place where the prey will be buried along with a wasp egg.

unmolested. They renew the numbers of their particular brood and species.

The value of such conformity is shown by those few periodical cicadas that fail to fit the program. Any that emerge a year early or a year late are gobbled up before they can reproduce the hereditary error.

As we think about the almost endless array of inherited strategies that insects use to escape from being eaten, by avoiding notice or repelling attack, we realize that few rely on a single system. The underwing moth that blends so imperceptibly in color and pattern and by obliteration of shadow on the

The owl butterfly has large markings on its rear wings, which are supposedly so similar to an owl's yellow eyes that they will frighten birds away. This specimen was photographed in Guatemala.

bark of a tree may startle a predator that comes close by raising the forewings to expose the rear pair, perhaps with bright pink stripes in a bold design. At the last moment before being seized, the underwing moth will leap into rapid zigzag flight that most birds have trouble following. Even if the pursuer catches the moth by its wings, there is still a chance that the overlapping scales on the insect's wings will come off, allowing the wing itself to slip free from the bird's beak. How the insect escapes is less important than that it live long enough to reproduce its kind.

5

Messages in Mate-Finding

SOLITUDE SEEMS SCARCELY A NORMAL CONDITION FOR ANY ANI-
mal no matter how much we admire its ability to cope unaided.
In the 1740s Charles Bonnet, a French biologist, discovered
virginal generation—parthenogenesis—of female aphids.
Each one could mature all by herself in less than three weeks,
and produce as many as sixty young females to repeat the
process. Bonnet also claimed that his microscope let him see
within the body of the pregnant mother aphid not only the
young she was about to bring forth, but within those young the
young of the next generation—the theory of preformation,
and one no longer subscribed to.

We know now that aphids and scale insects do reproduce
unmated all through periods of fine weather. Their partheno-
genetic development makes them explode in population, with
no time or energy spent on anything else than feeding, but
when cold weather or continuing drought arrive, these crea-

tures produce males and females that mate and produce fertile eggs. Only these resistant eggs have a chance to survive the adversities of a harsh environment.

The customary behavior of these particular insects in mid-summer gives little hint that such small animals conceal impressive capabilities, which often appear in the sexual behavior of male and female together. A courtship gesture, a territorial threat, a sharing act, all have meaning when received and interpreted by some other member of the same species. Delicate nuances confer on the interchange many of the features of a private language. Prey species benefit by keeping their communications sophisticated, peculiar, and inconspicuous. Predators gain by detecting signals and by eavesdropping, then pouncing on the signaler or the signalee or both.

Recently scientists have invented instruments that extend our own abilities to hear, see, smell, and explore the messages that insects send and receive. More challenging to interpret are touches and expressive nudges, such as those during a mating dance, because they are so delicate and quick.

RIPPLES IN THE WATER FILM

The quiet surface of pond and stream is more than an elastic support on which water striders may scull or hop. It is a medium for telegraphic communication by these insects, which make ripples of their own that are understood by other striders of their kind. The signals made by Australian striders caught the attention of R. Stimson Wilcox of Purdue University. He recorded their messages on magnetic tape and played back these signals to free striders while tapping out imitation messages. The insects responded to each type of communication with fresh ripples of their appropriate type or moved quickly in directions that convinced Wilcox the insect had interpreted correctly.

Male striders apprise females of sexual readiness. Females respond when they choose to. A male generates his signal either while standing free on the water surface or while grasping some floating or fixed object with his forelegs. First he stands and "calls" by shaking the surface film with up-and-down movements of his rowing legs—the middle pair. He sends out seven to fifteen concentric waves at an average rate of seventeen to twenty-nine per second. Each "call" begins faster (twenty-three to twenty-nine waves per second), then stabilizes at an intermediate rate, and ends with one or two low-frequency waves (ten to seventeen per second). Wilcox could recognize no individuality in these courtship messages, for each male varied as much in his wave output as one male did from any other.

A female strider within range of these relatively vigorous wavelets may respond first by sculling toward the caller. As she approaches him, she is likely to produce signals of her own. She generates low-amplitude vibrations of the water film by vertical movements of her forelegs. The male responds in the same way, as soon as he detects the approaching female. However, he receives no invitation to mate with her until she indicates her willingness by crossing one long leg over his, or grasping and pulling briefly with her forefeet at a middle or hind leg of the male.

A male that clearly gains a female's approval will back away from the object to which he has been clinging. She moves forward and grasps it for support. He mounts her, and they mate for about a minute. Dismounting, he backs away while facing her, and produces postmating signals indistinguishable from those the two insects used for courtship. Usually the female frees herself from the support, then turns about and grasps it with her hind legs, while using her abdominal tip to excavate a cavity. The male stands by until she has laid her eggs in the hole. Females seem to leave eggs only on objects from which a male has just been signaling.

It is easy to conclude that the male is protecting his mate, for he will shake the water film aggressively if another male approaches. The stay-away vibration pattern has a lower rate (nine to thirteen wavelets per second) or a distinctly higher one (twenty-three to thirty wavelets per second). A male may warn off another while free on the surface, where fights often occur and last up to several minutes. More often, the signal is a territorial sign from a male that has found a good site but, as yet, no female to come there and lay her eggs. He holds his place, shakes the water film, and rushes forth to do battle if an intruding male approaches to within three or four inches.

Our own photographs of water striders in groups on streams in New England sometimes show one individual surrounded by the concentric ripples of a vigorous message he has just produced. The delicacy of this means of communication among insects on the surface film almost defies human comprehension. Konrad Wiese of the Technical High School in Darmstadt, West Germany, confirms that a single vibratory wavelet can alert half the insects atop the water if the infinitesimal movement shifts the surface film up and down (or down and up) a mere thirty-nine millionths of an inch. The spectrum available for such communication ranges from a rate of repetition of 2.5 wavelets per second at the low end to 150 per second at the high end. The amount of energy in a wavelet becomes the limiting factor at the low rate, whereas the physical characteristics of the tension between water molecules sets the upper boundary.

THE OLDEST LOVE SONGS

The makers of most insect music began more than 300 million years ago to produce significant sounds and detect them. The calls of the hopping orthopterans, whether grasshoppers, crickets, or katydids, antedated by many millions of years the earliest voices of amphibians. More aeons passed before the songs of birds and the vocalizations of mammals were added to these sounds in the wild. Yet the insects have no voice, no combination of "tongue and lung," as Aristotle noted while examining carefully all the living things around his home. Unlike the vertebrates that came later, the insects scrape out their signals, or drum them vigorously, or just buzz their wings in ways that send sound in all directions simultaneously—while the flier proceeds, if at all, in only one.

The stridulations from members of insect orders that develop without a pupal stage all depend upon special structural features and highly purposeful behavior. Many of the performers will continue in an apparently normal way despite moderate confinement. A cicada produces an incredible sound while held in the fingers. Yet it is no toy, running on a hidden battery. It is the lineal descendant of an old-style insect, one now so familiar that many people know it as a "harvest fly" because its loud singing begins on hot summer days when the hay crop is ready to be hauled to the barn. Some of the cicada's kin among the other sucking bugs make softer sounds and detect airborne vibrations with great sensitivity.

Most sounds from insects that lack a pupal stage are made by the males. At close range the message may repel other males as a territorial warning. At greater distances, it attracts females, which approach the males and engage in courtship before mating. In a different sense, it indicates the readiness of the male for a sexual encounter. His song fades away as he transfers sperm, and rises in volume later when he has renewed his supply of sex cells.

The loudness of adult cicadas is real. In the tropics, where most of the fifteen hundred different kinds are native, their insistent calls ring forth like an alarm clock at dawn, and provide background sound from the high canopy of trees in the rain forest until the sun sets again. In years when they are abundant, virtually every cicada over much of the continental United States will synchronize his sound output with all others in his vicinity. The intensity of the combined calls rises into the range between seventy and eighty decibels. Heard at a distance of twenty feet, such as on the ground below the insects, the noise level corresponds to that measured from heavy traffic on a city street.

The unmuffled sound from each cicada emanates from special organs that occupy most of the basal segment of his abdomen. At the surface, this organ is roofed by an elastic membrane suggesting a drum. The cicada uses internal muscles to make his membranes vibrate, either vigorously or gently, generally in rhythms that carry a specific message. Its phrasing becomes evident in recordings made with a sound spectrograph, but meaningful details are scarcely audible to human ears.

Before refined equipment was available, the Rumford Professor of Physics at Harvard University, George W. Pierce, built a primitive analyzer with which to explore the insect sounds he heard around his summer cottage near Franklin, New Hampshire. His findings were surprising. The northeastern cicada known as the dogday harvestfly, for example, produces a carrier wave at high pitch, then modulates it 180 to 360 times per second. The outcome is "somewhat like the sound made by a circular saw going through a board"—a powerful, raucous scream. A smaller cicada of the north woods sustains a still higher note, pulsed irregularly about 335 times per second.

During the middle 1950s, Richard D. Alexander and Thomas E. Moore from the University of Michigan turned directional microphones connected to better recording equipment toward males of the famous periodical cicada in adjacent Ohio. These insects make an appearance there only at intervals of seventeen years, and the opportunity was too good to miss. To everyone's astonishment, two different patterns of sound overlapped, although the choristers appeared virtually identical. Each has a black body bordered in orange-red, with pale wings and feet.

One song in the insistent duet came from males that were, on the average, slightly larger than those that produced a second type of call. Entomologists identified particularly large individuals as "the" periodical cicada, *Magicicada septendecim,* and small specimens as a questionable variety, perhaps a separate species that could be called *M. cassinii.*

Incredibly, the songs of all individuals of both kinds synchronize in the same woodland. And when each phrase ends, many males fly to new sites as though they were playing a game of musical chairs. Early and late in the afternoon, a male may call half a dozen times before moving. In the midst of the combined chorusing, the moves may follow every phrase. The insect may shift only a few inches, or several yards. Sometimes the flutter of wings from those that have not yet found a new perch conceals the beginning of the next outpouring of sound.

When a female settles nearby, each male has a distinctive courtship call. He takes no chance of repelling her with loud sound, which carries easily for half a mile in all directions. At close range he invites her to respond to a gentle message. It is a slight buzz, repeated about five times in ten seconds, if he is *M. septendecim.* He produces paired noises with a far greater range of frequencies, about seven times in ten seconds, if he is *M. cassinii.* These distinctions clearly hold immense importance to a female of either kind.

Sound production and associated behaviors follow totally

different patterns among the ground crickets and tree crickets, the long-horned and short-horned grasshoppers that chirp and trill, buzz and scratch in distinctive ways, some in bright sun and others in darkness. Although the numbers of these orthopterans may be great, so too is the number of animals that eat them. Every cricket or grasshopper ceases sound production as soon as it detects a disturbance in its vicinity. A protest squawk, such as that of a captured cicada, seems out of the question; a cricket or grasshopper will try to wrench itself free, even at the expense of losing a leg or two, without uttering a sound.

Undisturbed males of this ancient order of leaping orthopterans seem ready almost everywhere on land in season to make their appeal over the sound waves. Least musical are the scratchy or rattly summer messages in daylight from short-horned grasshoppers. The chirps and trills of long-horned grasshoppers and crickets during sunny hours or darkness are more birdlike.

The short-horned grasshoppers show great uniformity in the way they hear, but considerable variation in the manner with which they produce sounds. All of them have earlike organs, one on each side of the first abdominal segment, with external membranes that remind us of the eardrums of a frog. Yet some of these insects, such as the common Carolina locust, appear completely silent while on the ground or feeding. They attract attention to themselves while in display flight, showing off the yellow border around their black hind wings while snapping the narrow front pair of wings against the rear pair.

Grasshoppers that stridulate while standing still or walking about produce the sound by rubbing the thigh (femur) portion of their hind legs against their first pair of wings, which are covers closed over the folded rear wings. But here heredity makes a major distinction. Some of these insects (members of particular genera) bear a close-set row of pegs or a hard ridge with many fine teeth (a "file") on the inside of each thigh, which they scratch against an equally hard crosswise ridge without teeth (a "scraper") on each wing cover. Members of other genera have the scraper on the inner surface of each femur, and the pegs or the file on a vein of each wing cover. In either arrangement, the insect frets file against scraper to create a chirp, just as a person can make a sound by scraping a fingernail along the tips of the teeth on a pocket comb.

The readiness of short-horned grasshoppers (locusts) to scratch out their distinctive calls in full daylight helps scientists who attempt to correlate the behavior of the insect with the sound he makes. Motion pictures of stridulating males can be compared closely with recordings of their messages. Whether the file is on the thigh or the wing makes little apparent difference, but even with these simple structures the insect can offer a surprising range of sound output. He has several ways to adjust his call: by changing the rate of movement of thigh against wing or the frequency with which he repeats his message, or by not scraping the full length of his file, or by pressing together the parts of his sound generator firmly and thereby softening his call. He can make a loud sound in attracting a mate or in threatening another male that is competing for territory, then transform the sound to a courtship invitation as he circles a female of his kind.

The sprinkled locust of New England adds fine details in the message by vibrating his wings while his thighs move past. The male of this species scratches the 104 teeth in each file on his femur against the scraper on his forewing in such a way as to produce a zipping sound. The high note lies close to A-sharp in the eighth octave above middle C. Probably this is the resonant frequency of the forewing itself, whenever it is caused to vibrate by sweeping contact between scraper and file. Each zip, however, is a train of eight or nine pulses at this frequency, because the file is not kept against the scraper throughout the quick swing of the thigh. The wing vibrates about seventy times per second, causing the scraper to strike some teeth, then miss some before striking again to create another pulse of sound.

Long-horned grasshoppers and crickets differ completely from the short-horns in the way they hear and produce sound. Each male matures with forewings that overlap somewhat at the base, the right wing held just above the left. This brings a file on the under surface of the right wing against a scraper on the upper surface of the left wing. In slow motion, the fretting animal appears to be shrugging its shoulders. At actual speed, the movements are too quick for the eye to follow.

No doubt the maker can hear his own chirp, thanks to small earlike organs that are flush with the surface of his front legs, just beyond the knees.

Two South African scientists, C. J. B. Smit and A. I. Reyneke, stimulated new studies of these stridulating insects by reporting that quite immature short-horned grasshoppers show stridulatory movements of their legs. For a young cricket or long-horned grasshopper, sound production is out of the question until after the final molt. The silent female also possesses a special inheritance. Only the signals from a male of her kind will induce her to approach and accept his procreative participation.

Since 1960, scientists have been able to overcome the inbuilt obstacle to interspecies crosses by perfecting the delicate technique of artificial insemination. Australian field crickets of two kinds have become favorite subjects for experimental study, because the hybrid eggs do develop and the resulting hatchlings can grow to maturity. Male hybrids tested in the Langmuir Laboratory of Cornell University by a team of scientists led by Ronald R. Hoy chirped in patterns showing an obvious compromise between the different songs of their parental species.

Surprisingly, the hybrid males reveal in their stridulated messages whether their male parent was *Teleogryllus oceanicus* or *T. commodus,* the two field crickets from the Australian community. Moreover, the difference between the two hybrid songs matters to the hybrid sisters. A wild male of *T. oceanicus* buzzes twice, the first buzz longer and followed after a short interval by a briefer one. A hybrid male whose contrived father was *T. oceanicus* produces a short buzz followed at intervals by three shorter ones. A hybrid male carrying the inheritance provided by *T. commodus* sperm begins with a somewhat shorter buzz, followed at similar intervals by six or seven much shorter buzzes.

The experimenters gave the hybrid females a choice between recorded songs played simultaneously from loudspeakers, one to the right, one to the left. Each could stay where she was, or approach either speaker. Seldom did she hesitate when offered the sound of a hybrid male and that of either male parent; she chose the hybrid. And when the two unlike hybrid male calls were offered, she crept two or three times as readily to whichever speaker played the call of a male whose male parent was *T. oceanicus* and mother *T. commodus.* Her own an-cestry in this reciprocal cross made no difference. Often she not only walked directly to the loudspeaker from which the recorded call of her choice came so clearly, but she climbed upon it, as though trying to get closer still to the insect she could hear through the high-fidelity equipment. Hoy and his colleagues could account for the sound-producing behavior of the hybrid males and the sound-preferring behavior of the hybrid females only by concluding that members of both sexes inherit a common feature in their nerve circuits, coupling them for effective communication in bringing together poten-tial mates.

Locating the control center for the sound-related behavior should be much easier in insects than in birds, because the nervous pathways contain far fewer cells. The chirping "crick-et-on-the-hearth" *(Acheta domestica),* a European insect that has followed humankind to most continents and found crumbs and refuge from outdoor cold in houses, has its largest nerve cells in an abdominal ganglion. These giant cells are excited when sound vibrations are detected by special receptors in the body wall, where it presses against the ground, as well as by receptive hairs located on a pair of short appendages (cerci) at the tip of the abdomen. The cricket has access to its heritage of chirp recognition in these particular cells, which process for it the strictly physical sensations.

Could an insect be taught to respond to a different message if exposed to the sound during formative ages? At the State University of New York in Albany, R. K. Murphey and S. G. Matsumoto offered the same kind of house crickets a chance to learn a tone pulse not too different from the species chirp. By rearing captives with at least the same mother, the experi-menters sought some uniformity among individuals that would be exposed to the repeated sound, and those that would be reared in comparative silence. Males were tested between two and five weeks after they had attained maturity. The scien-tists inserted electrical probes into the giant nerve cells and recorded the response there to tone pulses "of constant inten-sity (80 db) and 12 different frequencies arranged in random order, with 5 minutes between trials." How did these guardian cells react to human attempts to change the code? They resisted attempts to excite them with sounds no house cricket would make, and showed much less readiness to ignore vibrations at the frequency used by free members of their species. By comparison with cricket siblings in the untrained group, these males revealed an increased sensitivity for their own kind of songs. If they learned anything from experience before the test, it was how to ignore as noise any message inconsistent with their genetic heritage.

A pale green katydid, twitching her long threadlike antennae while she creeps through the dark or flies in a woodland at night, is unlikely to notice much difference whether the male she hears summoning a mate makes three basso chirps ("Ka-ty-did") or four ("Ka-ty-did-n't"). Our chance of seeing the insect himself is poor, since he blends so perfectly with the foliage where he perches on some tree.

A tree cricket is more likely to be trilling from a low shrub,

A pale green katydid resting on a green leaf at night need move only his front wings at the base to produce his loud scratchy call. He listens to other katydids with earlike organs just below the knee of each front leg.

his thin, flat forewings elevated above his back while he frets them together to sustain a tone like elfin bells. If we are extremely quiet and look for him with the aid of a flashlight, being careful not to alarm him by shining the beam too directly toward his eyes, we may watch while he produces his sound. His wings quiver far too fast for the human eye to follow, but he often continues for minutes at a time. Generally he adjusts his movements to tune his pitch until it matches almost perfectly that of others of his kind in adjacent shrubbery.

Occasionally we see a female tree cricket creeping slowly toward a male. Her long antennae whip abruptly from one direction to another. Her slender, pale green wings contrast

The male tree cricket raises his front wings above his back and frets one against the other to produce the trilling sound, which invites a female to nibble on a gland exposed on his back while his wings are up; he mates while she nibbles.

with the broad, flatter ones he holds elevated. She approaches her chosen mate and nibbles from a gland far forward on his back. It is a lure, exposed by the position of his sound-producing wings. While she nibbles, he shifts his position ever so little. He reaches his sex organs to hers, and transfers to her a sac of sperm suspension. Her organs accept his gift, and press its contents into her body, where the sperm remain alive and available to fertilize the next eggs she lays. We regard this as the most critical stage in the maneuver for, if the pair are disturbed, she ceases to nibble from his gland, and turns instead to eat whatever remains exposed of the sac of sperm suspension.

The female tree cricket has slender wings and makes no audible sound as she approaches a male; her long slender antennae reach far in advance, but he makes no move when she touches him.

WHIRS AND CLICKS
OF LATER INSECTS

As we think about the whole array of audible communication among insects, we realize how remarkably little has developed among the relative newcomers to the scene. The insects that transform from larva to adult during a pupal stage, dating back a mere 100 million years or so into the later days of the extinct dinosaurs, show a modest complexity in their calls and impress few people. Some beetles squeak. Some scaley-wings (the moths and butterflies) click. Some two-winged flies make recognizable notes with their buzzing wings, as do many of the various bees and wasps. Yet only a few of these squeaks and whirs and clicks seem important to the species that produce them.

Charles Darwin noticed clicking butterflies in Brazil: "Several times when a pair, probably male and female, were chasing each other on an irregular course, they passed within a few yards of me; and I distinctly heard a clicking noise, similar to that produced by a toothed wheel passing under a spring catch. The noise was continued at short intervals, and could be distinguished at about twenty yards' distance." Members of more than a single butterfly genus in the American tropics are known as "crackers" or "clickers" because of the sharp sounds they make so freely. They seem to possess no special organs with which to generate the vibrations, although a few scientists suspect that adjacent areas on the wings can serve in this way. It is quite possible that inaudible ultrasonic clicks and buzzes may yet be discovered from these insects.

Generally the male clicker butterfly perches head downward on a tree trunk, with his wings opened wide and pressed against the smooth bark to conceal his shadow and blend with his background. A passing butterfly of the same general brown coloration, or even a tossed stone three inches in diameter, may induce the resting male to fly in quick pursuit.

The fruitfly, barely more than a sixteenth of an inch long, shows that complicated behavior is made possible by having body organs serve multiple functions. The love song of the male, for example, is a product of the same wings and muscles he uses in flight. He vibrates his flight muscles at normal speed and extends his wing to adjust the pitch according to the hereditary pattern that is correct for his species. The female fly has no ears to pick up his sound. She uses parts of her extremely brief antennae instead, without sacrificing their ability to inform her about odors or to monitor air movements. The slightest amount of dirt on her antennae deafens her until she grooms herself clean again. In so small a body she has no backup systems in case her sensors fail.

The tiny fruitflies, which make courtship songs in many dialects, might best be studied in Hawaii, where almost a thousand different species are known—half of the total number for *Drosophila* in the world. Actual research on the male love songs has been done chiefly under indoor climatic conditions at Edinburgh University, where H. C. Bennet-Clark and A. W. Ewing looked into the various species that are used in genetics. In all species, the male produces his faint message as he approaches a female that attracts him. It is a one-to-one statement of interest, and clearly intended to elicit a response from her. Yet it goes further: it helps the female learn whether he is of her particular species. Each kind of male courts in his own dialect. All females share a single language, but it consists of only one word: a long and very loud buzz, the meaning of which to any male is "No!"

The male of the famous *Drosophila melanogaster,* for example, extends one wing and vibrates it while running around a potential mate. He produces a sound a few notes below middle C on the piano scale, and repeats this brief message twenty-nine times a second, for a second or two. A male of the closely similar *D. simulans* performs in the same manner, pitching his message in the same key, but he repeats himself only twenty times a second. Any female *Drosophila* worthy of motherhood must know when to say no, and when to stand still until a male of her own kind has mated with her, not just once but several times in rapid succession.

Virility in these little flies might be measured by the speed with which a male gets cooperation from a female of his kind, how many times he mates with her while she is still responding to his love song, and the number of offspring he fathers through this behavior. At the University of Birmingham, D. W. Fulker found that 3.6 minutes of courtship, followed by six matings in about 4 minutes, could provide for 340 fertile eggs. Slower males have fewer young to carry on their heritage.

The whining note from the wings of a female mosquito that is circling, ready to alight and probe for a blood meal, stimulates us through our ears to grow tense with apprehension. When the sound ceases, we try to deduce where she has settled, then swat the spot, hoping she will never fly again. A male mosquito would be more critical, since he can tell, within a few wing beats per second, whether her flight sound is the one that females of his species produce.

No one really recognized this ability in male mosquitoes (which suck only plant juices) until World War II. Then vast numbers of virile suitors of one mosquito species came rushing from the New Jersey marshes to some new electrical transformers that had been installed near Princeton. The transformers hummed with a seductive sound the males could not resist. Day or night they flew straight for the cooling fins on the electrical devices and died of heatstroke. Their bodies accumulated on the ground below, and had to be swept away in countless millions. So far as we can learn, this way to lure males to their death has never been used in an attempt to control mosquito populations.

CHEMICAL COMMUNICATIONS

Odors too fine to be detected by human noses or sensitive instruments provide important cues for insects. Crickets apparently reveal their sex this way, by releasing odorous secretions along the whole length of their long antennae. Either a male or a female that crosses antennae with another member of its kind can tell instantly whether those slender organs belong to a member of the same sex or the opposite.

That odorous secretions should influence insect behavior seems logical. Any animal of small size and terrestrial habits could do no better than show special sensitivity to small molecules. These tend to be the volatile ones emanating from plants and animal bodies. At extremely short range, they afford marvelous cues from the environment. They inform the insect about fragrant food, sites suitable for deposition of eggs, potential mates, and perhaps dangerously close predators and parasites. Through evolution, appropriate glands enlarged to produce odorous substances for chemical communication, on schedules that became increasingly specific for each species.

Many odorous materials used in insect communication are now known to perform a variety of functions: magnets capable of attracting large numbers of a species, regardless of sex; lures to bring members of opposite sexes together; territorial signals; trail-marking substances; and summons for colonial insects to join in an attack or a defense operation. A whole array of compounds keeps members of a colony integrated in their activities.

The word "pheromone" has become widely adopted to refer to any exciting chemical substance that serves as a messenger, acting at a distance from its producer on other members of the same species. The word appeared first in *Nature* in 1959, coined by P. Karlson and M. Lüscher. But the idea that one animal might perfume the air in such a way as to influence profoundly the behavior of another can be traced back to Aristotle. In his encyclopedic *Historia Animalium,* he mentioned that the name "zephyria" was given any infertile egg such as "young birds, as fowls and geese have been observed to lay . . . without any sexual intercourse." People believed the birds responded to the stimulating "winds in the spring" (the zephyrs).

Actual examples of insect pheromones remained unknown until less than a century ago. True, the eminent German biologist Karl T. E. von Siebold did suggest in 1837 that a female insect might possibly emit distinctive odors that would attract a male of her own kind. Equally probable, he said, would be male odors that could stimulate the female toward copulatory contact. But who could measure odors, or know one scent from another if it differed only slightly? It was easier to ignore suspected olfactory cues and investigate, instead, visible signals between insects or any sounds audible to the human ear.

The French naturalist Jean Henri Fabre was the first to explore an olfactory system. He discovered it after a female of Europe's largest moth (the great peacock) emerged in his home laboratory on the morning of May 6, 1875. Fabre put her in a cage made of wire screen. About nine o'clock that evening, his household was invaded by male "suitors" of the same kind,

The fragrance of the female Promethia *moth (with less feathery antennae) brought the male (here a larger individual) upwind to her side, where courtship could continue and mating follow. (Photo courtesy Esther Heacock)*

entering open windows on all sides of the house. "Coming from every direction and apprised I know not how," he wrote, "here are forty lovers eager to pay their respects to the marriageable bride born this morning amid the mysteries of my study."

Before Fabre's caged prisoner died, she attracted 150 males on the eight consecutive evenings. Yet, in that part of France, cocoons of this moth seem scarce because the old almond trees (on which the caterpillars live) are few. The males must have come "from afar, from very far, within a radius of perhaps a mile and a half or more. How did they know of what was happening in my study?" At the time, Fabre noted that the males needed their feathery antennae to find the female. Among sixteen individuals that he deprived of these organs, not one joined the unaltered males in flying to the female's cage as soon as all were freed within the same room.

Four years later, as Fabre described so well in *The Life of the Caterpillar,* he proved without a doubt that some scent too subtle for any members of his human family to detect provided the potent lure. A female enclosed in any kind of hermetically sealed container—wood, metal, or glass—could summon no males. A narrow crack sufficed to let her scent escape and attract potential mates, unless the gap was covered by an inch or two of cotton. Fabre found males of a related moth, the lesser peacock, responding in the same way to a female of their kind. But each male came only to the lure of a female of his own species.

A German scientist at the Max Planck Institute of Biochemistry, in Munich, Adolf F. J. Butenandt, became curious about the magic substance that could affect moths so impressively. He dreamed of being able to synthesize the chemical compound responsible for this behavior, hoping to achieve this step for moths whose caterpillars were pests of agricultural crops, in order to lessen their economic impact without affecting other species. In the 1930s, before being honored by the Nobel Prize committee for his earlier work on human sex hormones and cortisone, Butenandt decided to examine carefully the lure of the silkworm moth. This insect seemed ideal, for it was raised in great numbers in many parts of the world for the production of silk. Its caterpillars mature quickly in captivity, and the moths themselves are large enough to handle with bare fingers. Moreover, its close relatives (which might be similar chemically) include several whose caterpillars are destructive pests.

Butenandt found that an empty box in which a female silkworm moth has spent a short time will attract great numbers of fluttering males. They come to a paper cone that has been smeared with a crude extract of the lure glands from 100 female silkworm moths, while ignoring a live, unaltered, sexually receptive female a few feet away. He reasoned that, with a suitable formulation, he could entice a whole population of males from a pest species to overlook potential mates, or totally confuse the males as they sought to follow odor trails to receptive females, or draw the males into traps where they could be exterminated. Butenandt did not foresee one other major benefit now recognized from his research. Using the sex lure of a pest species, entomologists can capture and identify the first males to emerge in an outbreak year, in time to apply other control measures and minimize reproduction. Knowing when and where to act is often half the battle.

From the Far East, Butenandt imported a million silk-moth pupae. Aided by a team of coworkers, he dissected out the appropriate glands from 310,000 virgin females. Extracting and reextracting, he used live male moths to show him by their behavior which component contained the effective substance. Many months of work went into refining the material until it was pure and potent. By then the biochemists had a mere twelve milligrams (423 millionths of an ounce) of it to analyze. But a bare quadrillionth of a gram elicited an ecstatic response from a male silkworm moth. The scientists named the compound bombykol and discovered that each molecule of it contained sixteen atoms of carbon, thirty hydrogen atoms, and one oxygen atom, and had two important twists in its three-dimensional structure. Synthetic imitations without these twists arouse the male silkworm moth scarcely more than pure air.

We like the German word *Lockstoff* that Butenandt and his coworkers use for this powerful substance. It unlocks the mate-seeking activity of the males of this one species without affecting males of other kinds. The lock that this single molecular key fits so well consists of special receptor cells in the feathery antennae of the males. These cells fire off a nerve impulse to the moth's brain if they pick up just one molecule apiece. So hypersensitive are the cells in a male moth that those in his two antennae send about sixteen hundred false alarms per second, with no stimulation at all. The moth ignores this many. But if two hundred more nerve impulses per second are added, from cells that have picked up two hundred molecules of the key substance, the male turns and takes to his wings. He flies upwind to the attractive source of this distinctive perfume. For him, it is the perfect aphrodisiac. Curiously, a human nose gets only a faint scent, like leather, from a concentrated solution.

Bombykol was the first sex attractant to be identified chemically, as Butenandt and his collaborators pointed out when announcing their discovery in 1959. Sex lures from female insects of more than two hundred different kinds have now been studied with equal care. All prove to be relatively simple compounds, which a competent chemist can synthesize in several forms. Each will have the same fundamental constitution, but differ in the shape of the three-dimensional molecule. The receptors on the insect's antennae respond fully only when the molecular shape is exactly right. Even then, the principal lure may have to be accompanied by one or more supplementary scents at close range, to assure the male that he is approaching a potential mate of his own species.

The sex lures produced by female moths all prove to be mixtures of two or more volatile substances. Yet their chemical simplicity is as subtle as that of the two plain keys needed to open a safety-deposit box. They must be presented at the correct place and time. They serve solely when two individuals who can identify themselves in other ways are ready to use the keys in interaction for a special service.

Often the female remains quiet. As Fabre wrote, she "seems indifferent to what is going on." Yet, often, she has only a few days of life left. She must respond at the first touch, the first certain fragrance, and perhaps the sight of the correct male to receive his sperm and fertilize her eggs. She must place them where the hatchlings will find food upon emergence, all before her time is up.

This bias toward having the males move and assemble, while the female waits, arises from the logistics of the species. When, for example, a female gypsy moth hatches from her cocoon, she bulges with about four hundred heavy eggs. They weigh her down, making flight impossible even though her wings are somewhat larger than those the lighter males use so effectively. She waits awhile for some male to respond to her lure. If none comes, she may lay most of her load before she dies. But the eggs will all be infertile, wasted from the standpoint of her species. Some other form of life will benefit from the nourishment the eggs contain.

Entomologists feel justified in investing large amounts of time and money in research aimed at identifying and synthesizing the lure substances of particular pest species. In this way they hope to deflect the normal behavior of the insects in directions that will protect food crops and stored products for humankind. Nor have the moths, beetles, and other pests any way to develop an immunity to this invasion of their private chemistry. They have no substitute way to find mates if they ignore the natural lures that have served their species so well for countless generations. If inheritable change permits a slight alteration of the chemical message, like a change in the secret code, the chemist can notice this and adjust the synthetic recipe.

The amount of perfume that a female insect releases can be infinitesimal and still effective in summoning a mate. A large female may manage with as little as a small one. This presents real obstacles for the biochemists who seek to identify the important ingredients. "It's like looking for a single, sharp odor in a chemical haystack," says Ted Underhill of the Canadian Research Council. Better methods than were originally used to identify the lure of the silkworm moth have been developed, first to extract enough material for the chemist to analyze and then to discover what part of the molecule the male finds particularly attractive.

Surprisingly, the big cockroach that is native to the southeastern United States (known there as the "palmetto bug") proved extraordinarily secretive about her lure. The males detected it and came running. But the biochemists got nowhere by extracting the glands they suspected of producing the special fragrance. Finally, they tried another approach. They enclosed 500 adult female cockroaches in a thirty-gallon can for several days, and gently moved air through the container, past the insects, then into a trap chilled with dry ice. The lure condensed in the trap as a small pool of oily yellow liquid, and a male cockroach could smell a drop of it a mile away.

Males too have their chemical signals. A male cockroach that finds a female of his kind, either by following her lure or by accident, releases a fragrant secretion that persuades her to allow him to mate. "Seducin" is the delightful name given this substance by L. M. Roth, who discovered the male's behavior at the U.S. Army Laboratories in Natick, Massachusetts.

Admittedly, the male of almost any species is challenged if the female he finds is free to move, perhaps as free as a female butterfly. In some way he must get her attention away from food or other distracting influence into a sexual trend. For his own satisfaction and the good of the species, he may have to be devious. Some male butterflies achieve this by turning odorous glands inside out from the posterior end of the abdomen as they fly just ahead of a female of their kind. The fragrance from the glands has a flowerlike scent to a human nose, and seems to suggest a source of nectar to the female butterfly. Commonly she extends her tongue, at least partway, as though ready to sip.

Some of the bark beetles seem to have the lure system reversed. It is the males that emit the scented substance in the first place. Each summons females to some chosen site on a dead or dying tree. Females respond to the odor and assemble. Each becomes pregnant in turn, and cooperates with the male in chewing a passageway into the bark to the place where she will lay her eggs. While she busies herself in raising a family, he renews his chemical invitation to more females, wherever they may be.

The famous crickets-on-the-hearth have an odorous communication system based on one attractant substance—which both sexes apply to their favorite perches and both detect—and two repellents. The attractant is fairly stable and continually attracts the crickets to certain vantage points. The repellent produced by the male, and spread on any surface where he stands awhile, helps keep other males away. Since the repellent is more volatile than the attractant, it ceases to be the dominant scent about twenty-four hours after a male has ceased to rest on any particular site. Females apply their different repellent, which has much the same characteristics; apparently it keeps the population spread out, and reduces competition for resources such as hideaways and food. Neither sex pays much attention to the repellent of the other.

The great mystery is how even the most obvious system of pheromones came into operation, for the fossil record has yet to yield a hint. Perhaps sexual differences in the odor of body wastes led gradually to other roles and statistically significant benefits to a species. The suggestion that the lure of the female, at least, might be derived rather simply from chemical compounds in her plant food, has led to investigation but has yet to yield hard evidence.

The source of the female's lure remains as mysterious as ever, but the monogamy in many insects, following a single successful mating, is now known to be due to a contribution from the male. The male heliconiid butterfly, as he mates, drenches his female with a volatile substance that repels other males and swamps her lure. Lawrence E. Gilbert of the University of Texas, who discovered this explanation for the female's subsequent behavior, calls the material an "antiaphrodisiac." He suspects that it benefits the species by saving time and energy, both in the mated female and in males that might otherwise try to court her. The effect lingers in these long-lived insects for weeks or even months.

The antiaphrodisiac, which has an odor that a human nose could mistake for the sharp fragrance of witch hazel, is probably carried on a pair of "stink clubs" near the end of the female's body. They are so positioned that they dip, during mating, into a pair of pouches the male possesses. His pouches are lined with glandular tissue and have no other suspected role in his particular life-style.

BRIGHT SIGNALS IN THE NIGHT

Each June brings us one sign of the seasons we especially enjoy. We record the date, and experience a quiet satisfaction to see the world of nature keeping time so well. The event cannot be forecast exactly, but we feel confident it will occur on June 10 or on some subsequent night before June 22. Call it an end to springtime and the approach of the summer solstice. The signal we look for is bright dots and dashes toward the end of twilight, as fireflies begin their communications of the year.

Across much of the Northern Hemisphere, "lightning bugs" have been summer magic since prehistoric times. The Japanese admired the fragile beetles that provide the display, and continued the old story from their folklore that a firefly was the spirit of a fallen samurai warrior, returned to see how his country was faring. Parents caught fireflies to release inside the mosquito netting over their children's sleeping mats, so that the youngsters would drift off into dreamland watching these insect stars winking harmlessly in the darkness so close overhead. Today, in much of Japan, the custom is vanishing— as are the wild fireflies. Polluted air, soil, and water in the industrialized nation, together with the expansion of pavement, have all but eliminated the places where fireflies used to grow to blinking maturity.

We hear that the Edogawa district of Tokyo spends $15,000 in public funds each year to breed and raise a batch of fireflies so that once every summer the children of the neighborhood can experience what their parents and grandparents remember. Hisashi Abe, who is in charge of the project, opens the box after dark on a July evening that is announced only locally, so that visitors from elsewhere will not arrive in uncontrollable crowds. Out come hundreds of fireflies, blinking their lights against the darkness.

Scientists have devoted countless hours trying to discover the firefly's secrets. No one really questions the suitability of glowing bodies and winking lights as signal systems to bring males and females together for mating after dark. Yet the fact that immature stages of these insects glow steadily suggests that the firefly glows because it cannot help it. The real accomplishment of the insect as an adult may be to turn off the display, to change a steady light into a variable or winking one, and then make use of it in communication.

The glowworm of literature is the wingless female of beetles common in moist, temperate parts of Eurasia. The female is larger than the male and is about three-quarters of an inch long. Just the last few segments of her body emit light when the glowworm glows. She compensates for her small size and for the fact that her luminous organ is more under her abdomen than exposed on top of it. She creeps up a plant stem or a grass blade until she is well above the ground. Then, glowing, she rocks from side to side, directing her light in both directions. Her "belly dance" speeds up the longer she waits. People notice glowworms beckoning a mate after dark late in June and through July.

The flying males are less conspicuous, each barely half an inch long, and possess only a modest beacon. They have excellent eyes, however, and look downward as they cruise about. Upon seeing the glow from a female, the male spirals through the night as best he can and makes a landing close by. Since beetles are mostly rather clumsy fliers and the general illumination is too little for him to see the low vegetation, his first approach may be up the wrong stem. Or he may stumble against the support to which the female clings. A shaken glowworm or one that is touched turns off her light immediately.

A male keeps his faint one lit while he hunts for her, and afterward if he finds her. We might predict his move if her light goes out before he arrives. He may just wait until she turns on her glow again, as though she had forgotten his blunder and perhaps noticed his ready light. But he might also see another glowworm at full brilliance somewhat farther away, take to his wings, and try his fortune elsewhere. Since males and females of this species are about equally numerous, his chances are good.

The smaller of the two European glowworms and their flying consorts make an appearance in late July. Males outnumber females five to one, and seem particularly energetic. Often a male will spend many minutes courting one of the larger glowworms and being rejected by her, before abandoning this endeavor and trying elsewhere. Many a male must go mateless, unable to find his glowworm. But with so many males, each female has a fair chance to be discovered.

The seeming coyness of European glowworms, which turn off their lights so readily, cannot be all gain. Professor Friedrich Schaller of the Zoological Institute, Braunschweig Technical High School, finds that a male of either species can become distracted right in the middle of mating if another glowworm turns on her light within his view. He may disconnect himself and seek out the new attraction. Worse still, many glowworms fail to glow unless the night is dark. A faint beam from a flashlight suffices to turn them off, and in the vicinity of large cities, the night sky is often too bright from street lamps for them to glow. Even where soil conditions offer suitable hunting grounds for immature fireflies of this type, they simply disappear, their potential prey uneaten. Perhaps efforts to conserve electric energy will cause a resurgence of Europe's glowworms.

Eastern North America has a true glowworm too, but you have to be lucky to find her, or to recognize her once you do. Like the larva she resembles, her body is wingless and bears about a dozen bright beacons along each side. When they shine forth, she is the most brilliant of fireflies in the country. Yet she hides where the larvae do, under stones and fallen logs. The male, which lacks light organs, must locate his mates by sight and scent. He scarcely resembles a beetle, for his front wings are too tiny to conceal the flying pair. He rarely folds them, as other beetles do when they settle. The only one we have encountered, a male almost three-quarters of an inch long, kept flying or running about, using his big eyes and his prominently comb-shaped antennae. We lost sight of that one male without a glimpse of his mate's luminous display.

Most fireflies in the North Temperate Zone wink their lights, unless they are captured and held. Then they turn on their display and keep it going. Even after a dozen fireflies have been caught and swallowed by a toad, they shine for a while —brightly enough to show right through the toad's stomach and skin. The normal routine, however, begins with a message in coded flashes from a searching male as he flies above a field or between low shrubs. One particular kind of firefly we all meet every summer *(Photuris pennsylvanica)* produces a bright greenish-yellow signal consisting of a brief blip followed by a prolonged streak. Ham radio friends insist this is the symbol for the letter A in Morse code, with the sender slow about lifting his finger from the key switch on the "daah" of "dit daah." The firefly repeats this signal at intervals of about $7\frac{1}{2}$ seconds, and makes each message last almost three seconds.

Other kinds of fireflies in the same genus have different call letters. A single blip of light (the Morse code for letter E) is repeated at regular intervals every second by *P. frontalis,* every two seconds by *P. hebes,* every four seconds by *P. cinctipennis.* A double blip, rather too generously spaced to be a good Morse signal for the letter I, is the visible signature of *P. fairchildi* males, which repeat it every $5\frac{1}{2}$ seconds. A streak of light (the Morse for T) is the sign of *P. caerulucens,* which turns on his luminous organ every four seconds and lets it shine a full second before turning it off again. The distinctive *P. lucicrescens* behaves as though he had a dimmer switch to play with; he turns up the intensity of his display from zero to full brightness in just under two seconds, then turns it off completely for $5\frac{1}{2}$ seconds before repeating his signal. The well-named *P. tremulans* seems to the human eye to produce a quavering light every $7\frac{1}{2}$ seconds, although actually it consists of a dozen flashes in the same second each time. Differences in the color of the light emitted seem less important to the insects than the pattern in each signal, the intervals, and any variations in intensity.

The different firefly signals are invitations by a flying male for a female of his species to recognize him and to invite his close attention. His eyes scan unblinkingly, forward and downward and to both sides, watching for a luminous answer. The nature of her flash is less critical than its timing. He will ignore it if it comes too soon or too late to be the appropriate message.

He flashes. She lights up at the correct number of milliseconds later. He executes a perfect U-turn in the dark and flashes again. She answers, once more as a clear response and with the delay that is correct for his species. He flies closer and sends another signal. She answers. The visible two-part harmony continues until he alights beside her. Then touch and chemical interchanges can confirm the mutual suitability and readiness of the two to mate.

Quite a number of different firefly females fatten themselves on males that they lure down to certain doom by communicating in the code of their victims. James E. Lloyd, an entomologist at the University of Florida in Gainesville, performed some experiments in the field. First he located females of a predatory species (*Photuris versicolor*—call her "V") that he suspected of undercover activity. He found them by their flashed answers to passing males, and to penlight simulations of male firefly signals. Next, he watched V females to see what they would do as free males of their own and other species flew by, signaling in the darkness. Properly, a V female responds with an emphatic blip a second after she sees the triple flash of her own kind of male. What would she do if stimulated by passing suitors of species M or C or T? Could she imitate the response of an M female, a C female, and a T female, and react correctly to a V male as well? Two of the suspected V females reacted with just such versatility, and summoned both T males and C males with adequate imitations. A third V female invited a passing M male with a good counterfeit flash even though she was still devouring a male of a still different species she had lured within reach. One V female captured the twelfth M male she answered. Another moved to a new location after her false flashes had failed with twenty C males in succession. From the new site she continued, and caught C male number twenty-one there. Occasionally a V female stopped responding to any male, or flew away. Lloyd could not tell whether she gave up because of increasing age, perhaps the condition of her ovaries, or from a long sequence of unsuccessful captures, or from prolonged exposure to flashes of males other than her own kind.

Eleven females that Lloyd tested with artificial signals responded appropriately to the flash pattern of M males, then switched promptly to behavior needed to simulate a T female. On seven occasions, the V females interrupted Lloyd's experiment, ignoring the signals he provided to answer appropriately quite different patterns emitted by passing males of other species. Lloyd called their behavior "aggressive mimicry," and the fraudulent signals the products of "femmes fatales." Their imitations proved "quite effective, and females seldom answered more than ten males without catching one" for dinner.

Male fireflies varied too in their behavior when Lloyd tested them with short flashes of light from a small lamp bulb at ground level. Only about one C male in ten would come to the bulb when it was set to glow in short blips, spaced like those of the female of this species. Yet one C male that a wolf spider caught continued to emit his rhythmic pattern. Two additional C males "were attracted to the flashes of the captive, and were also seized by the spider."

Winking o˙ glowing lights do bring together the adults of unlike sex reliably for almost two thousand different kinds of fireflies. A great majority live in the humid tropics. Those of Jamaica have been collected and classified with special diligence. It wasn't until the 1960s that experts attempted systematically to discover whether a person could learn to recognize each species of flying male by his pattern of illumination. Are these communication signals as distinctive to the human eye as birdcalls are to the ear of a field ornithologist?

A group of curious experts from various research centers in Maryland decided to find out. They acquired suitable equipment, made arrangements through the director of the Institute of Jamaica, and set up operations on the island. For fireflies of certain types, they centered their study on an open area of

Castle Hill northwest of Long Bay. For others they chose a remote section of road threading through the foothills of the John Crow Mountains. At each location, one person aimed the sensitive equipment at every male firefly that came along in a fairly straight flight path, while a second investigator operated the recording system. Other participants hovered nearby with insect nets, ready to capture the firefly as soon as the recordings were complete. The insect went into a vial, numbered to correspond with the record chart, for later identification.

The scientists learned quickly. They distinguished readily among the luminous insects those that produced a long-continued glow, which usually fluctuated in brightness; almost certainly these were members of genus *Diphotus*. Other types emitted single flashes shorter than a tenth of a second, but at regular intervals of two or more seconds. Still others sent out one or more twinkles every few seconds, a "twinkle" consisting of from four to more than twenty short flashes in quick succession. Often the human eye could not resolve the separate peaks of a "twinkle" that showed up reliably on the recording, and that a female firefly might well detect.

With experience, the investigators found themselves distinguishing species in the dark. Yet the records raised unexpected questions. Do fireflies migrate up and down a mountain slope, producing unlike flash patterns according to altitude or to age? Does the difference between the timing of light output by a male on the wing and the same individual while he rests on a leaf provide a different message to a female? How does one interpret the change in a firefly's signal when he is excited by proximity to a potential mate, as compared to his excitement when he finds himself confined in a small vial? The communications of a firefly proved much more variable and complex than anyone anticipated.

In southern Asia, from India through the East Indies to the northern tropical tips of Australia, many of the male fireflies belong to species with the special habit of flashing in unison. For months, they provide rhythmic bursts of light from trees, at intervals ranging from a half second to one second, depending on the kind of firefly and the temperature. The display may continue for hours each night.

One of the earliest explorers to report such synchronized

behavior was a Dutch physician, Engelbert Kaempfer, who traveled downriver from Bangkok to the sea in 1680. "The glowworms . . . represent another shew, which settle on some Trees, like a fiery cloud, with this surprising circumstance, that a whole swarm of these insects, having taken possession of one Tree, and spread themselves over its branches, sometimes hide their Light all at once, and a moment after make it appear again with the utmost regularity and exactness."

Other observers have reported this phenomenon, and sought to explain it during the 300 years since Kaempfer's discovery. The descendants of the fireflies he described still come to mangrove trees along the river near Bangkok. The trees cannot be approached easily on foot, for they grow in deep mud, but from a small boat the visitor can see the fireflies at close range and distinguish the thousands of tiny lights.

John Buck and Elizabeth Buck, who traveled from Bethesda, Maryland, with special equipment to analyze the display, reported: "Each time we saw this hurrying, soundless, hypnotic, enduring performance it impressed us anew as uniquely different from any behavior we had ever seen." Each flash, they found, is double: a faint blip followed almost immediately by a brighter one. And each male rests on his own mangrove leaf, defending it from the approach of any other male. His flash is a warning, a sign of occupancy.

How can every male blend his display into one pattern? Unless he is a newcomer to the tree, still searching for an unoccupied leaf, or he is disturbed in some way, he flashes within a fiftieth of a second of all the others. So long as the night air holds to about 82° F. (28° C.), the fireflies maintain their rhythm within 0.006 second of the average 0.560 second between flashes—just slightly slower than twice per second. Because this is too fast for any firefly to detect the beginning of his neighbor's flash and make his own light organ emit in time, he must adjust his own output in relation to the latest flash he sees, and continually monitor his behavior according to whether his output is a trifle early or late by comparison with the others. Experiments with captive fireflies have proved that he can adjust to match a rhythmic pattern from a flashlight bulb, even if it is considerably slower or faster than that normal in a tree full of insects.

The Bucks credit the female firefly with rewarding only those males that keep perfect rhythm, while ignoring any individual that attempts to express his own identity, to capture attention, or to "cheat" on the system by flashing early or late.

The total production of light from so many small insects becomes phenomenal. Yet, for individual luminosity, no firefly seems able to maintain his or her output of illumination as spectacularly as certain click beetles of warm climates. For a moment, a firefly may outshine the click beetle. But the winner keeps glowing brightly from each side of its thorax, above and below, in a behavior that was once counted on to light a surgeon's work when the lighting system failed during the construction of the Panama Canal. Our own experiences convince us that a few luminous click beetles in a clear bottle furnish enough illumination for a person to read the Miami newspaper during a blackout. The only requirement is that the surgery or the reading occur on a night soon after the particular beetles attain maturity, and in a place where they can be collected after dark.

Pyrophorus, the "fire carrier," is the generic name for most of these luminous click beetles. Our first experience with them came unexpectedly, on our very first camping trip together. We had put up our tent at sunset in a remote and unfamiliar bit of Texas desert. We closed ourselves in the automobile to avoid the chill outside and turned on the light to write up our notes. Outside, coyotes were now calling in dozens of voices. We turned out the car light, and prepared for the quick trip to the security of the canvas tent. But paired lights remained, pressed against the sloping windshield. We flicked the switch and looked more closely. Just slender click beetles an inch in length stood there, each with its twin luminous spots. When we switched off the light again, we saw others of these insects flying, and a few perched on an adjacent cactus plant. The beetles too could turn their lights off and on.

Possibly the largest species of *Pyrophorus* showed us the role of its luminous organs in social interaction above the treetops in a steep valley on the Pacific slope of Ecuador. The display of living fireworks began as evening twilight faded, and lasted less than twenty minutes. Each fire beetle flew at high speed, circling, rising, falling amid a group of eight or nine. Several at a time blazed their paired lights above and below, brilliant orange against the darkness. At times they flew close enough that we could distinguish two flaring spots on the near side of the beetle because it was sideways to us, and one spot on the opposite side due to the banking angle of its curving turn.

We tried to lure one of these amazing insects within reach, using a penlight flashlamp, but the ruse did not succeed. Months later, we discovered in the scientific literature concerning fire beetles the fact that they can count. *Two* lights an inch or so apart, turned on for a few seconds and then off for a similar length of time, will attract the flying *Pyrophorus.* One light cannot be another click beetle, and the live insects ignore it.

The chemistry of beetle display, whether by click beetle or firefly, is now more widely known. Three substances go into the recipe for light production: a special enzyme (luciferase) to facilitate the reaction; the specific substance (luciferin) that releases the light energy as the reaction simplifies and destroys the compound; oxygen; and adenosine triphosphate (ATP), an energy storage compound contained in every living cell, and essential in the insect to make luciferin available for use. The luciferase and the precursor of luciferin can be extracted from the organs of fireflies. Purified, they provide the most sensitive material yet discovered with which to test for minute amounts of ATP, and have even been used to test for traces of life on Mars.

On earth, firefly extracts are used routinely for tests as varied as those used in the study of heart disease and muscular dystrophy, urinary infections, the effectiveness of antibiotics, of waste-water treatment methods, and for early diagnosis of hypothermia in swine—a condition that costs the pork industry nearly $300 million annually. In fact, conservationists feel growing concern over the increased exploitation of fireflies and the continued inability of chemists to synthesize the essential ingredients for imitating insect production of light. In many states and countries, children receive a penny apiece to collect fireflies by the millions so that the mysterious ingredients can be extracted from their luminous organs. How long will the wild resource continue, and will it shrink through extermination before anyone learns the details of the extraordinary communication system and the still more baffling chemistry? We need to know far more about the ecological roles these insects play, and how light serves them as they produce it at so many stages in their lives.

At a time when energy for human activities, including electric power, must be used wisely lest the supply fall short of legitimate demands, the luciferin-luciferase system presents a special challenge to technology. The reaction releases more than 75 percent of its energy in the form of light at wavelengths visible to the eyes of fireflies and humans alike, and less than 25 percent in the form of useless heat. By comparison, the efficiency of a fluorescent lamp is nearer the ratio of 30 percent light and 70 percent heat, and the incandescent bulb emits only 10 percent or less in the visible spectrum—the remainder is total loss. As any child who holds a firefly knows, the light organ is as cool as the rest of the insect. "Cold light" is a correct description of bioluminescence.

BODY LANGUAGE

Dancing and courtship go together as symbols of sexuality in much of the animal kingdom. Impromptu as the bobbing heads, gesturing arms, swaying hips, and shuffling or leaping feet in a primitive human tribal rite, or precisely stylized as a classical ballet, these movements mime a message. The non-human counterparts, whether lordly elk, whooping crane, scaly lizard, finny fish, grasshopper, or butterfly, repeat inherited patterns of behavior. They stimulate excitement as well as demonstrate it. Performed correctly in suitable context, they generally induce a coy and reluctant female of the species to permit liberties that she would otherwise repulse.

Nothing else an insect does suggests play so much as a courtship dance. It may all be necessary to get the potential partners into the right stage of mutual acceptance and suitable position. Often the interacting pair posture and pirouette, frisk about and take protracted rests as though in no hurry to shorten the duration between reciprocal discovery and sexual consummation. The antics of mating time contradict the general impression we get that insects show impressive dedication, a fervent fondness for feeding undisturbed, and immediate irritability or readiness to escape if interrupted in the routines they follow from birth to death.

Courtship dances await discovery almost anywhere on land or over fresh water. Niko Tinbergen at Oxford University tells of courtship behavior among European grayling butterflies. Day after day he watched and tested these insects in the sandy country of the central Netherlands. He distinguished males by slight differences in color and by their habit of settling on "regular observation posts, either . . . on or very little above the ground." Whenever another butterfly flits near, the male grayling at once dashes after it. Surprisingly, they pursue butterflies of other species, wasps, dragonflies, grasshoppers, dung beetles, even their own shadows! Equally enticing is a paper butterfly towed by a thin thread from a stick about three feet long. Fluttering movements prove more important than shape, size, or color. A rhythmic flutter is one form of dance.

The female grayling might agree with Jessica, in Shakespeare's *Merchant of Venice,* that ". . . love is blind, and lovers cannot see/The pretty follies that they themselves commit." Yet, we notice, some female graylings are so coy that they respond to the pursuit of a male by flying off as fast as they can. Not until a male catches up with one will she settle to the ground. The male quickly alights close by and walks toward the female. She recognizes that he is her kind, but has not yet accepted him. She may simply walk away, flapping her wings vigorously, signaling that she has already mated and will not accept another suitor. Males abandon wing flappers, and renew their search. But if the female sits still, she is a virgin, awaiting the next move by the male. He walks around to face her, holding his wings folded together tightly above his back, just as hers are. Now that she can see him at close range, he extends his antennae straight to each side and waves them continuously, causing their tips to follow small circular or semicircular patterns. She shows some interest by raising her own antennae, turning toward the male's wings. Next comes his move; he brings his forewings forward, close to her antennae, then separates and closes the leading edges while keeping the remainder of his wings tightly together. After a few seconds of this strange wing-fanning, he suddenly parts his forewings, moves them farther toward her, and clasps her antennae between them. Even then she may refuse him, leaping into flight as soon as his wings touch the knobs on her antennae. If she accepts him, she will let him walk quickly around behind her, slightly to one side, and bend his abdomen to achieve sexual union.

Rarely is this pas de deux so uninterrupted. Two grayling butterflies that we followed for almost half an hour on the Swedish island of Öland repeated the early parts of this routine first one place, then another, because of a gusty wind. It swept one insect or both off the exposed soil before the male could press the female's antennae between his forewings. They finally succeeded in mating, but the wind kept swiveling them about. So far as we could see, the male made little attempt to stand on his own feet while his mate struggled to do so. He swung in circles around her, his sexual parts and hers holding the two together. Often the wind pressed their closed wings flat against the ground, as horizontal as the sails on a small boat in danger of capsizing. The two maintained their union for nearly ten minutes. Later he might watch for another mate. She, we were confident, would just flutter her wings at the next male that caught up with her.

Maintaining contact until the fruitful finale of the courtship dance must be at least as trying for other attractive butterflies. We marvel at the persistence they show many times each summer in our own vegetable garden. Perhaps a male cabbage butterfly discovers a female sipping nectar from a flower. He fairly twitches with excitement, alighting near her wherever she goes. For a while she may ignore him. At last she allows him to walk so close behind her that he can join himself to her. Does he relax completely then? The twosome give that appearance, for she still flies from place to place, waving her wings, while he just bumps along behind her. His wings press together above his back. His sexual linkage holds too firmly to jar loose. The male butterfly shows no awareness of other males that flit about his mate, or that many of her short flights are moves to avoid them. He retains his position, inconvenient as it may appear, until he has transferred all the life-giving semen she can take or he can give on this occasion.

Exploring high mountain slopes, whether in western America from New Mexico northward or in the Alps and elsewhere across Eurasia, we encounter butterflies with special challenges in courtship. They live only near tree line and above, as tail-less high-altitude relatives of lowland swallowtails. Named *Parnassius* to remind people of Mount Parnassus, where the Greek gods held lofty court, these insects are twice as big as a cabbage butterfly. They cope well with high mountain winds that must tear at their ivory-colored wings, each marked with a pattern of black dots and a few red ones. A *Parnassius* is equally recognizable and distinctive in flight, or clinging to low vegetation, or finding refuge in the lee of some rock.

The mating ardor of the *Parnassius* butterfly is cooled by dense clouds that cast shade on a sunny day, even if they shed no cold rain, snow, or hail. When the sun shines again and the wind relents, we see these intriguing butterflies dancing about, seeking mates. We have not learned to tell a female *Parnassius* from a male unless she has already accepted a suitor. Thereafter she carries over her abdominal tip a conspicuous membrane that resembles a petal from a flower. It is attached there by her mate, as a product of a peculiar gland in his abdominal tip. Entomologists call the membrane a "sphragis" and note with some amusement that its presence does not stop males from paying attention to mated females, although it may discourage them sooner. The membrane completely shields her genital openings, and works like the chastity belts of the Middle Ages.

The little springtail males of the water film demonstrate that this flexible surface is like a water bed—an exciting place upon which to make love. They run about, nudging the larger females from all sides, until a cooperative mate is found. She lowers her antennae. The male crosses his antennae with hers, and suddenly locks himself to her by pincer action, where two adjacent segments of his feelers bear appropriate, sturdy spines. The mechanism holds as firmly as the jaws of a plumber's pliers. As though startled, she lifts her antennae—and the male too. For days she may carry him about, he vertical, she horizontal, she feeding on pollen grains and other food particles caught on the water film, while he fasts. Not until she lowers her antennae again can he make another significant move. Yet he is ready, primed. He backs up a few steps

Parnassius butterflies lay their eggs on plants above the timberline on western mountains, where their caterpillars are in danger of freezing each night; a little at a time they attain full growth. The adults sometimes visit flowers at lower elevations, but few people recognize them as the tailless swallowtails of the heights.

at a time, towing her after him. He extrudes a droplet of mucus rich with sperm cells. Like a tiny pearl, it rests on the surface film without getting wet. Again the male walks backward, until he draws his mate over the droplet. Without hesitation, she scoops up the droplet into her reproductive tract, and some-how signals her success. He lets go of her antennae, and backs away, his service performed. Now it is her turn—to find a place for fertilized eggs, to start another generation of springtails on their chancy journey to the future. By then her mate has flipped off for a new adventure.

The courtship antics of springtails on land seem stodgy by comparison. The males lack antennal pincers, and generally mature at approximately equal size to the mates they seek. Nudges and odors provide the meaningful cues to insects of such minute size. Their eyes offer almost no picture of their world or their associates, and serve mostly to alert the springtail to any abrupt change in light intensity. Each individual springs into the air, abandoning courtship instantly, if a skunk raises the piece of rotting log or the stone under which the male and female have begun their courtship.

The mating games of bristletails may be still more ancient, although the fossil record has not yet revealed whether these wingless insects originated as early as or earlier than the springtails, in late Devonian times or before. Only the most insistent study, however, reveals the courtship details although the bristletails are larger, sometimes an inch and a half in overall dimensions. The most cosmopolitan bristletail—the silverfish—seems reluctant to show itself in any light bright enough for human vision. The equally common bristletail of wharves and pilings, of cracks in rocky coasts, runs and leaps in broad daylight, but proves unbelievably aware of any observer, whether curious scientist or hungry bird. Yet the inherited compulsion to find a member of the opposite sex and to mate, despite captivity, has sometimes been so strong that the insects ignored a human audience in dim light.

The game is similar, regardless of the site. The slightly smaller male bumps energetically into the female of his choice, waits a few moments, and bumps again. His rough treatment continues for so long that, at first, we wonder why she does not turn on him and retaliate in kind. The explanation is simple —males are rare. In fact, female bristletails are suspected of laying their single large eggs without fertilization most of the time, which develop by parthenogenesis (unfathered birth).

Fertilization is possible, even if rare, and is beneficial to the genes. Yet success for the male bristletail follows an odd course. After a prolonged episode of bumping a female and drumming on her body with his antennae, he spins a silken line from the tip of his abdomen and suspends a few droplets of sperm-charged mucus on this elastic fiber. One end of the line stays glued to the ground or a rock. The other he keeps taut, while he nudges the female around. The dance reveals no obvious pattern. Yet he maneuvers his mate as expertly as any tugboat captain pushing an ocean liner to the dock. The female moves around and back. She sidesteps this way and that, finally dancing right over that line with its glistening droplets. Without hesitation she takes them into her body and becomes pregnant. He can manage her this way in the dark, in a crevice, almost anywhere in the world.

Courtship in crevices, particularly for insects that avoid any light, affords favorable features for creatures of small size, as well as constraints we can recognize more readily. Odors linger longer where the winds blow past rather than through. A potential mate cannot so easily escape attentions before responding to them. Touch sensations, no doubt, become paramount, and taste offers extra communication with contact. We can scarcely learn how all these senses fit together into erogenous patterns for any insect, for our human experiences follow such a different model. The insect has nerve networks around each projecting bristle, but almost none below its firm, waterproof, continuous cuticle. Instead, its sense organs (except for the eyes) are concentrated on a pair of antennae, two pairs of extra appendages close to the mouth, three pairs of legs, and, in many instances, a further array at the base of the wings, and around the rear end of the segmented body.

Courtship in the open during daylight hours is safest for insects with huge eyes. Dragonflies and damselflies qualify superbly, and demonstrate their specialties above or close to fresh water. The largest male dragons patrol territories and drive off others of their sex. A female that comes along will not flee, as an intruding male would, and thereby earns a different reception. The male swoops low above her as she settles lightly on some conspicuous support. It may be a water lily pad, a large rock, a twig tip, or a tree trunk. Her jaws may still be chewing the mosquito she caught most recently. But her bulging compound eyes, which cover most of her head, inform her of his every movement as he hovers before her.

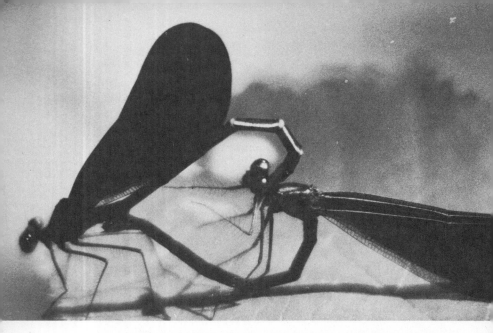

A mated pair of damselflies rests on a leaf close to a stream while she (right) reaches forward with her abdominal tip to pick up the packet of sperm he placed hours earlier in a special pocket below his anterior abdominal segments. Meanwhile, he clasps her by the neck with his abdominal forceps in the grip he must use to haul her out of the water after she has laid her eggs below the surface.

Often the female dragonfly, although surely satisfied that the courting male is of her own species, flits off into the territory of the next male of her kind. Possibly her acceptance of any mate depends upon internal events, such as the stage of development of her eggs. The antics of a courting male may stimulate the release of sex hormones that would accelerate this process, because the investment of food reserves in the eggs is too great to have them ripe for sperm entry before a male dragonfly is anywhere in sight.

The male continues courtship until the flights of the female are all short and roundabout. Inconspicuously, he reaches forward with his abdominal tip to deposit a packet of sperm from its point of origin into a peculiar reservoir and mating apparatus in the second and third segments of his long abdomen. (No other type of insect is so equipped.) Then, he dexterously alights just one body length ahead of the female. He stretches his slender abdomen backward above her head. Gently but firmly he grasps her slender neck in a terminal pair of fitted pincers, and vibrates his wings to become airborne. She follows his lead, providing her fair share of propulsion like the second rider on a tandem bicycle. Back and forth the pair will flit, the male sometimes engaging in a little hunting.

Meanwhile, the female awaits some internal cue that tells her when her eggs are ready. Then she curves her own abdominal tip far forward under the body of the male, to reach the sperm reservoir he filled earlier. For a time, as the paired insects fly about, they form a perfectly poised "mating wheel" that is unique in the animal kingdom.

Somehow the female dragon signals to the male when she has drained the sperm from his body into hers. Soon after it is time to thrust fertilized eggs one at a time beneath the surface film of suitable fresh water. The tandem dragons fly low, to hover momentarily while she strikes the water film with her tail tip and washes off an egg. Sometimes he lets her go, and she performs this gesture on her own. By dispersing her eggs widely, the dragonfly reduces the likelihood that the hatchlings (naiads) will have to compete with each other for prey. Each of the young will need to capture still-smaller creatures underwater for at least a year, or for as many as five. It will stalk through the aquatic world wherever it can catch insects, crustaceans, and worms as nourishment for growth.

Damselflies appear far too lightweight for such vigorous courtship. Yet in flight each is so exquisitely under control that it can hover close to a plant stem while it nibbles off one aphid after another. Each female watches the antics of an attentive male until she can accept him. He grasps her by the neck while the pair fly or rest in tandem. Generally he settles and raises his abdomen high up to aid her as she reaches forward to absorb his sexual products. Usually he holds her securely at the water's edge while she walks backward, dragging him with her, until she is immersed completely and only the forward three-fourths of his body, his wings, and his head remain in the air. Prodding with her abdominal tip into the plant stem that supports the two, she wedges each egg into position so far below the water film that a few days of dry weather will not normally lower the water level enough to let those eggs die of desiccation before they hatch.

The male damselfly inherits the responsibility for pulling his mate out of the water. He may need to flutter his wings in order to pull harder as her wet head, thorax, and wings come through the adherent surface film. She cannot escape the grip of the water without this help, yet if she takes her time, he may get distracted. Even though he can feel his mate's presence where his claspers hold her by the neck, he sees other females, as yet uncourted, winging past. They tempt him to let go and leave the immersed female to drown. We sometimes wonder if a rescuing finger to haul the female out of her wet predicament would make that much difference. She would probably never mate again, and, without a tending male, never place another egg where it could add to posterity.

The courtship pattern shown by two-winged flies of many kinds is "On with the dance!" Myriad beating wings create a faint, continuous buzz. For mosquitoes, midges, and various types of gnats, the favorite site and time is above the top of an elm, poplar, or other tree (including several kinds of conifers) in early evening. At that hour, vertical streams of moist warm air rise out of the canopy of foliage. The open metal structure that supports a major shortwave antenna or television array can also produce an ascending column of warmer air. Any of these sites may attract a swarm so dense that it resembles smoke. One 1912 report on the mosquitoes of North America, Central America, and the West Indies tells of townspeople who turned in a fire alarm when they saw such a cloud above their church steeple. The female insects detect the swarming males and fly in among them. The aerial dancers, rising and falling in vertical elliptic paths, guarantee a quick encounter. Rapidly joined, the pair sink from the swarm and consummate their nuptials on some firm surface, such as a leaf.

Only occasionally can anyone discover what visible features of a female insect excite a male of her species. One female will catch the attention of male after male, while another female is bypassed repeatedly. Even when we detect an obvious difference between two individual females, we must remember that the male insect's eyes are quite unlike ours. Although we see finer details at a distance than any insect, it sees features in the ultraviolet spectrum. To our eyes both sexes of the pale green, long-tailed luna moth appear equally pastel, but a male insect of this kind reflects so much more ultraviolet from sunlight that by day, to another luna moth, he is a blond, and she a dark brunette.

For butterflies, the critical detail that stimulates the male to display himself may be a glimpse of a female of the right kind or a whiff of her perfume. Seemingly, some insects use vision as the principal sense in mate-finding, whereas other insects rely on scent. A few years ago, Lincoln P. Brower of Amherst College in western Massachusetts found that among butterflies that rely principally on vision, the males all appear virtually identical in coloration, although they can cope with two or more different color patterns among acceptable mates. But where scent is the primary identifier, both males and females may come in several different garbs without causing confusion. In the tropics, where predators of many kinds watch constantly for prey, successive generations of the same kind of butterfly may appear quite unlike in hue and pattern according to the time of year, in order to mimic a whole series of unpalatable insects.

The armament of the famous hercules beetle of the American tropics serves as a weapon for dislodging another male, or as a fur-lined forceps for picking up and carrying off the female.

Distinctive differences in body form between male and female are well known in many insects. Yet only a relatively small number of instances have been linked to courtship behavior. The male European (and British) stag beetle has oversized jaws and will battle another male or pursue a small-jawed female in a suburban garden despite an attentive circle of human witnesses. His body and head may measure two inches long, his jaws extend another inch in front. He may be awkward, almost like a mechanical robot, as he uses those jaws to grapple with another of his kind. If his antagonist succeeds in overturning him, a male stag beetle often needs a minute or two to right himself. By then the victor probably has made off with the waiting female, or begun to mate with her.

Occasionally a trend in insect behavior, without much difference in body form, can be traced from one kind of creature to its relatives. We then suspect that the highest development of the behavior arose by easy stages, some of which were end points in other species. The eminent Harvard entomologist William Morton Wheeler delighted in one particular sequence of steps he arranged in order from his observations of small two-winged flies. One type, found from Mexico to Guyana, is a mere third of an inch in length, and its males swagger about on long legs over the surface of leaves, displaying iridescent blue bodies. Each male fly alternately extrudes from his mouth and sucks in again a large shiny globule, almost as though he were chewing bubble gum. If a female responds to this display, as though charmed by it, she shows her approval by hurrying over to share the globule. After a few minutes the droplet is gone, and the female may loiter briefly as though expecting another treat. The male often seizes this opportunity to mate with her, before she changes her attitude.

Some individual males prolong the female's docile state by daubing minute globules on the sides of her head. She then stands around, retrieving the sticky material with her forefeet and washing her face before departure, ignoring whatever the male is doing behind her back. Wheeler noted that many a female, after finishing the liquid refreshment and casting her mate aside, might promptly accept another partner, actuated (he suggested) "more by hunger than by lust."

Only an expert can distinguish between a micropezid fly, a

male of which offers droplets of lure to potential mates, and an empidid, which seeks still tinier insects as prey. Sometimes an empidid male mortally paralyzes his prey and flies about with it basketed between his legs until he locates a dancing swarm of females of his own kind. There he selects a mate and offers his prey as a gift. She takes it to a leaf, on which she can perch while she sucks the present, squeezing it to obtain every drop of liquid contents. The male mates with her as she stands there, busy with her bonbon. When the nourishment gives out, the female discards the shell and her attentive partner with a single gesture.

Closely related flies of similarly small size perform a more elaborate courtship ritual. In some, the male in flight picks up a minute stick, a flower anther, or other object, and spins around it a fluffy, balloonlike mass of silk from fine strands extruded by his salivary glands. He offers this toy to a female, who examines it intently and progressively unwraps the layers until she finds out what is hidden deep inside. This gives the male the time and cooperation he needs. But other males of the same genus produce the gift balloon of silk alone, and find it equally effective in courtship. Even a secondhand balloon will serve. Those that a female has dropped, "probably after pairing, are constantly picked up and used over again by other males."

The behavior of males in constructing nuptial balloons, which often enclose a morsel of food, may easily originate in patterns that serve self-preservation. A male (drone) honeybee has no such opportunity. Beekeepers know that disaster is guaranteed for any drone that succeeds in catching up with a young queen on her nuptial flight and inducing her to accept him. This high-speed courtship far above the ground suggests a dogfight between combat airplanes piloted by determined fliers of warring nations. Yet the goal is not to have one prospective partner destroy the other bee. It is to link the bodies of two individuals in prime condition, until the male contributes his total supply of sexual fuel for the queen's future egg-laying operations.

As soon as the young queen receives all the sperm cells that one suitor can supply, she breaks away from him forcefully and actually tears off his connecting organ, fatally wounding him while letting him go. He spirals to the ground to die in a few hours. Meanwhile the young queen ejects his organ from her receptive parts, and flies down toward the hive to attract another flight of drones. If she succeeds in mating with only two or three on nuptial flights the first day, she returns to the hive for the night and repeats her aerial maneuvers the following day. Not until the reservoirs in which she stores and nourishes the sperm cells are full will she settle down inside the hive for a lifetime—four or five years—of laying eggs. When she does so, every other drone in the hive is excluded by the workers. No more may they get food by soliciting contributions. In a few days they starve to death.

In New Zealand we believed for a few minutes that we had found a mosquito seeking a female to court by running about lightly atop the surface film of a small pool of water. We kept looking for a female underneath, and wondering whether he had lost her. He made short runs, stopping a few seconds to probe the water film, then hurried on another equal distance. Occasionally he stopped for longer periods, and this gave us the chance we needed to examine him more closely. Each stop of a minute or so preceded by seconds the emergence of a female from the pool. He seemed to be helping her emerge. Then we discovered the explanation of this strange behavior. His quest was for female pupae as they started to transform, to move from their floating "bullhead" pupal coverings to a winged adulthood in air. He did help pull each female from her pupal shell, but he mated with her at the same time. To us it seemed incredible how many virgins he could greet with this momentary courtship and brief insemination before running off to find another.

While the New Zealand mosquito actually gets to meet his mate, many a male bagworm moth sees merely part of the females he finds. He flies about in search of self-imprisoned mates. Each is wingless, completely enclosed in an elliptical shelter of silk and sticks. He settles on this covering, which she completed at the end of caterpillarhood and which hangs firmly from some twig or branch. (His pupal covering was much smaller because, as a caterpillar, he completed his feeding weeks of development at lesser size.) Any female bagworm that is ready for him has already softened the lower end of her

A bagworm carries its self-made case wherever it goes to feed; the case will become its cocoon. If the insect is a female, she will never leave this pupal shelter, but will lure a male with wings to fly to her and to mate. She will then lay her eggs inside the case she constructed while still a caterpillar.

cocoon. He dances around over the surface, and somehow detects a response. Gently he extends his own telescopelike abdomen through the opening in the female's cocoon, and mates with her. He withdraws, and takes to his wings again. He may have mated with a dozen females similarly before the first of them begins to lay her eggs inside her cocoon. She will shrivel and die there, never having seen her suitor or the world he flew in, unless dimly through her juvenile, caterpillar eyes. Her hatchling caterpillars will emerge from the bag, to feed when the tree next puts out a fresh crop of leaves.

Certain female bagworms native to the Southern Hemisphere construct, as caterpillars, the largest of these strange shelters. Some in South America and Africa attain four inches in length. They impede any form of courtship we could regard as complementary dancing, but protect the female and her eggs from many a hungry bird. One bag of record size came to us in the hands of a missionary to Nigeria, who wanted to know what it was. We ourselves encountered native people in Swaziland who had a personal use for these insect containers. The Swazis were on their way to a wedding dance, all accoutered in traditional style. Around each bare human shank, a snug harness held a dozen or more bagworm bags. Every bag had been freed of its insect inhabitant, dried, and provided with a pebble. When the human dancer pranced, the pebbles shook and rattled. The bridegroom and the bride seemed especially adept.

6

Extending the Heritage

ALMOST ANYWHERE THAT PEOPLE LIVE, YOU CAN FIND A HOUSEFLY
and give it a moment's consideration. Its life began minutes
after its parents mated, for fertilization occurs while the egg
is on its way to the outside world from the ovary inside the
female. The fertile egg was laid with perhaps 125 others in the
batch, and took about ten hours to hatch, if the weather was
warm, and up to three days if the surroundings were cool. The
maggot needed only five days to attain full growth in a suitable
situation, and another five inside its last larval skin (puparium)
to totally transform its body to that of the adult. Then it burst
free through one caplike end of its covering and crawled forth.
The time clock began silently to measure the days until the
inner heritage would end the creature's life.

Under the most ideal conditions, the muscles of the male's
wings begin to weaken by his fifth or sixth day of adulthood.
His wings fray and tatter progressively. By the seventeenth day

—the end of a normal lifespan for his sex—almost no male houseflies have enough wing membrane and wing muscles left to fly. Only one percent of the males will live to be forty days old, measured from the beginning of emergence and adulthood. All of his flying about, finding nourishment and mates, courting and passing on his heritage must be accomplished in that time.

A female housefly inherits greater durability. Her average adult lifespan is twenty-seven days, which gives her slightly more than two weeks to accept mates and lay eggs. Until her fifteenth day, her wings will show little tattering. If she is among the privileged one percent with great longevity, she will live fifty-five days and be able to fly until the last one. By then she should have laid six hundred eggs—and be a grandmother! Ten generations in a year is easy for a housefly, but no male of the species has a chance to meet—let alone recognize—an adult among his offspring.

Sips of water can help a housefly of either sex to live its normal lifespan. Sips of milk can extend the survival of a female, for she can digest and use the energy from milk sugar (lactose). The male lacks this ability. His inheritance gives him a brief period in which to serve his species through reproduction, and then ends it forever.

The roles of adult insects sharply vary according to sex, and the difference in demands often leads to genetic disparities. The male can contribute his share to the diversity among the offspring through the sexual process with only a modest investment in energy or time. For him to participate in tending the eggs or young is rare. It is the female who loads the eggs with yolk for their embryonic development, and who seeks a place where she can leave them protected. Often she travels great distances between the place where she mates and the one where her young (and his) are to take their chances in surviving. She may even stay to guard and nourish the next generation. Yet she earns no leisurely retirement in which to enjoy her world after these vital tasks for her species have been completed.

Most adult female insects are either genetically capable of feeding themselves or must rely on the nourishment they acquired as larvae. The mayfly is in the latter category, and must

Eggs of a bug develop where their mother attached them to a leaf, on the plant her scent receptors and sense of touch led her to choose for her young. The dark eyes of the unhatched bugs can be seen through the transparent eggshells, from which we can see that all are turned in the same direction.

rapidly mate and lay her eggs. For species that can digest carbohydrates, such as those found in the sugars of nectar or honeydew, there is more time to reproduce. The females that can digest proteins have an even greater advantage, however, since their ovaries can continue to produce eggs almost indefinitely.

The heliconiid butterflies of the New World tropics play the game both ways. Members of the genus *Dryas,* such as the julia butterflies whose caterpillars feed on the foliage of passionflower, have ovaries that shrink at about the same pace as the insects use up their larval food reserves. Members of the genus *Heliconius,* which includes the common zebra butterfly of Florida's southern tip, extend their lives with pollen. They collect the golden dust and keep it moist, perhaps with nectar, until each pollen grain begins to grow. This process liberates amino acids, which the butterfly then absorbs. The continual and theoretically unlimited source of dietary nitrogen allows the butterfly to survive, producing an almost endless number of eggs, for as long as six months. A *Heliconius* that is given only nectar will merely place the eggs developing in her body at the time of adulthood, and die in a month or less.

Pollen has proven to be an important source of nourishment for many adult insects that visit flowers. Honeybees carry quantities of it to their hive and store it in special cells as "bee bread." A large reserve of this orange-colored, somewhat bitter material and of honey is sensed somehow by worker bees, and induces some of them to partially digest so much bee bread that their salivary glands begin secreting "royal jelly." The rich jelly is fed to certain larvae in brood cells of unusual size and location, spurring the growth of the larvae to mature as virgin queens instead of workers. The royal jelly is also fed to the queen, keeping her alive and active. Like magical water from a Fountain of Youth, it allows her to survive for as much as five years of busy egg-laying, scores of months longer than any of the worker honeybees with the same heritage.

The genes concealed within each insect define a proportioning in the adult life of each female between investment of energy in eggs for the benefit of potential young, and survival of the parent until she has completed her work. Any inherited strategy that calls for her to obtain more food than her body contains when she reaches adulthood seems a tradeoff. It lessens the space inside her abdomen that can be given over to reproductive functions by requiring the continued presence there of digestive organs. For carnivores or blood-suckers, however, the digestive system can be more compact than one required to get comparable nourishment out of plant materials. With a few exceptions (such as queen honeybees), long-lived adult insects are predators or external parasites.

Praying mantises spend many weeks in late summer perched on goldenrods in our garden, waiting for extra nourishment to come their way. The daughter of a neighbor learned, when not yet four years old, to pick up full-grown, winged mantises and bring them to us. We showed her how to distinguish the slender male from the female with its broader abdomen, and warned her never to disturb a female that clung head downward on a plant stem. Such a mantis would be about to whip up a mass of tan froth and fertile eggs, and affix this creation to the stem. Ignored by most birds, the protected eggs would overwinter and not hatch until warmed by a few weeks of summer weather.

One afternoon our little friend came rushing to us. "Something awful's happening!" she cried, and led us to the scene. How do you explain to a child that a female mantis, while still copulating with the male she has accepted, will do her best to grasp his head and the forward portions of his body—and devour them? His heritage accommodates this form of cannibalism, for his abdomen continues to transfer sperm into her storage reservoir even after she has decapitated him and started to eat his thorax. Only after he contributes all he can (or she has space for) will he let her pull him free, and complete her meal. The nourishment from his body will go toward yolk in her next batch of eggs.

Jean Henri Fabre described this unseemly end to courtship in the praying mantis of southern Europe. Usually the male approaches a prospective mate quite slowly, gently, ready to run or fly if she makes a threatening gesture in his direction. No matter how cautiously he comes, she almost quivers with the appearance of suppressed excitement. Yet generally she makes no move, not even when he suddenly leaps and mounts her in the mating position. Not until a little later is he likely to become dessert.

This behavior of European praying mantises came with them to the New World when they were introduced as beneficial insects in Massachusetts. Kenneth Roeder of Tufts University had to go no farther than the nearest vacant lot to repeat most of Fabre's observations. Fabre was quite correct in noting that each female mantis normally accepts the attentions and sperm from a number of suitors, and that those that escape being eaten by her are the exception rather than the rule. Those that get caught, in Fabre's sympathetic words, "unite again in a more intimate fashion. If the poor fellow is loved by his lady as the vivifier of her ovaries, he is also loved as a piece of highly-flavored game."

Roeder noticed that the fate of the male depends mostly on how symmetrically he alights atop his mate. If his positioning is perfect, he may later leap away to safety, perhaps to approach another female on another day. If, however, he lands even slightly askew, she is able to reach out a grasping foreleg and get hold of his head. Roeder marveled that the nervous system in the male's abdomen can continue to coordinate his mating movements long after the large paired ganglia in his head (serving as a brain) and the connections farther back have been totally destroyed.

A blood meal for a female mosquito is very important, for without it, she may be able to lay only eight or ten eggs or, in many species, none at all. The nutritional supplement, gained in her own special way, lets her deposit as many as 200 eggs. Sometimes we watch closely as a mosquito settles light as a pillow feather on sensitive human skin. She may stand there, ready to rush away at the slightest disturbance, for a minute or more before continuing her search. Ordinarily she rests on all six feet, her wings silent and folded above her back. Should three of her legs be lost, she can sometimes still manage to hold her balance by using the other three as a tripod. An extreme case can occasionally cope with just one leg, the tip of the outstretched wing on the opposite side, and the end of the abdomen as support.

Perhaps fifty seconds are needed for an intact female mosquito to press her slender lancets through human skin, into a small blood vessel. She wets these mouthparts generously with saliva that acts as an anesthetic as well as a lubricant. The inward progress of the lancets is very smooth, in spite of the jigsawlike action of the tapered tips. Once in position, the mouthparts form the sides of a tube through which the insect can withdraw a load that often exceeds the dry weight of her own body. Sucking for as long as three minutes, she makes no move until stretch sensors in her abdomen report that she is near the bursting point. Then, in five seconds, out come her lancets. She regroups her mouthparts quickly for flight, spreads her wings, and takes off heavily on her reproductive mission. Rapidly her ovaries get the iron and proteins she has gained. None of this new nourishment, apparently, goes to the flight muscles. Their energy comes from sugars obtained in nectar, a resource from flowers of many kinds that she may need to replenish every thirty minutes by whatever route she travels.

A female flea takes a moderate meal as though unwilling to relinquish the narrow figure and light weight she needs to slip between the hairs of her host or leap for safety. Like the mosquito, the flea takes too little volume of blood to affect the host, and its worst effect may be to tickle the skin, inducing an itch, then a scratch, and perhaps a tear through which infection can enter. The flea must keep moving from one drilling site to the next, since the host may scratch the last place at any moment. Yet a small blood meal may be all that several fleas need to get them together, start an orgy, and later to send every pregnant female on a hunt for a crevice in which to leave her fertilized eggs. The hen flea warms to a mate before ever getting anything to eat, merely by landing on a suitable fowl or a warm substitute.

Because the role of the blood meal in the feeding and breeding habits of fleas seemed so well known in the late 1950s, quick success was anticipated by British scientists, who decided to raise rabbit fleas in large numbers. The insects could be freed in the countryside, to transmit the deadly disease known as "rabbit fever" (myxomatosis, due to a virus) and perhaps rid the United Kingdom of European rabbits. The fleas, however, failed to breed. They lived for month after month, through spring and summer, taking blood meals from caged rabbits before anyone suspected that the strange behavior might be linked to the practice of maintaining the rabbits in separate quarters, males apart from females.

An amateur naturalist, Miriam Rothschild, discovered the link, and with her assistant Bob Ford, demonstrated that the rabbit flea of Europe and the cottontail flea of North America need sexual stimulation in their meals. In the wild state, they get it from hormones in the blood of pregnant rabbits and newborn young. Only this chemical stimulant from their food induces the flea eggs to develop. The fleas of both sexes cluster without mating on the expectant mother and wait until she has given birth. Then the insects transfer to the newborn. They feed voraciously, mate repeatedly, and the female fleas drop off. They go to place their eggs among the litter in the rabbit nest, where the larval fleas will hatch and scavenge. Only later, when the young rabbits are about to be weaned, will the adult fleas return to the mother rabbit. It is she who will carry them farthest to fresh fields, and give each flea a sporting chance to change rabbits while the hosts are mating.

SECURITY FOR EGGS AND YOUNG

Each mother-to-be has her own program, her special ways to perform her procreative role. Occasionally the fossil record supplies a firm basis for modern ideas about the probable past history of insect behavior as it diversified and continues to evolve. Sawflies, for example, appeared about 200 million years ago, and are the earliest known members of the great order Hymenoptera, which also includes the ichneumonflies and other parasites, the solitary wasps and bees, and the eminently social insects. Ever since, they have shown the least specialization in their structure, the fewest changes in appearance, since those early days. They still have larvae so like a caterpillar in form that we can tell the difference only by looking underneath and noticing extra pairs of soft prolegs that support the hindmost three-fourths of the body. All along, female sawflies have possessed sawlike ovipositors, used to cut narrow furrows for their eggs into plant stems and leaves. The larva may feed in the open on foliage or crawl to a new location if it fails to find the appropriate leaves. Or the hatchling may inherit a more compact body form and little ability to travel on its own. In this case, the mother's routine with her ovipositor includes adding a glandular secretion to create a gall. The secretion causes unspecialized tissue to transform into specialized cells, and causes continued growth where ordinarily no further growth is called for. The epidermis of the plant must heal over the growth region and prevent infection, while underneath, the supportive tissues give characteristic shape to the gall, within which the insect larva will feed on the proliferating, and almost shapeless cells.

The initial stimulus from the mother insect is supplemented

subsequently by exudates from the developing egg, the hatching fluid, and the larva itself. These keep the plant tissue
undergoing cell division at a remarkably controlled rate. The
cells that the larva eats are replaced, and the shelter slowly
enlarges around the insect. Eventually—usually when the larva
attains full size—the growth of the gall ceases. By then a person who is familiar with galls can usually predict what kind of
insect is inside. In a sense, and due to the larva's secretory
contribution, the gall is as much a product of the insect as of
the plant. The genetic heritage of both species combines in the
interaction.

A single goldenrod stem, for example, can be deformed
halfway up by a spindle-shaped gall caused by a moth caterpillar, higher by a spherical gall induced by a gall fly, and at the
top by a loose community of gall midges. The moth caterpillar
and the gall-fly maggot scarcely affect the ability of the plant
to flower and produce abundant seeds. The gall midges, however, modify the reproductive portions of the goldenrod, converting it into a shaggy whorl of abbreviated greenery around
the central feeding region, with no flowers at all. Fortunately,
any one stalk of goldenrod is unlikely to have even one gall,
let alone two or three different types. Some evidence suggests,
in fact, that the presence of one gall inhibits development of
another, acting as a sort of vaccination. Willows and a few
other kinds of trees appear particularly susceptible to gall
insects of several kinds.

The eggs and hatchlings of predatory insects may be in
special danger from the mother that has just released the egg
or from the first young to emerge. This is probably why dragonflies wash off their eggs one at a time, and why so many
other predaceous mothers drop their eggs singly. Some of the
one-inch water bugs produce a whole raft of eggs at a time, in
a manner that appeals to many an advocate of women's libera-
tion. The female bug holds down the male after mating with
him, and cements her egg cluster to his back like kernels of
unpuffed brown rice, with one end up and the other cemented
to the body of their father. From then on until every egg has
hatched through its upper end, or until the pond threatens to
dry up, he moves about slowly and fights off any predator his
size that might threaten the burden on his back. At intervals,

he swims to the surface and rocks his body sideways, swishing the eggs into air and under water again, rinsing and aerating them thoroughly. Two weeks or more of this attention and the young water bugs emerge to hunt on their own. Pond water seeps then into the vacated eggshells and softens the cement, letting the male brush them loose against aquatic plants. If the pond gets too shallow, the cement hardens and cracks in the dry air, freeing the father of his progeny. In the normal course of events, however, the male is no sooner free of one batch of eggs than he seeks another female, who will reload him. This behavior is quite exceptional, however. It never appears among the larger water bugs of the same family, or in smaller members of this widespread group of predatory insects. They appear much more given to cannibalism both before and after brief episodes of courtship.

It is the hatchlings that provide the threat to one another among lacewing fly young, when the mother-to-be lays a series of eggs in rapid sequence. She gives each hatchling a chance to go off on its own, affixing every egg to a support. First she secretes from her abdominal tip a drop of cement on some stem or leaf. Next she stretches from this anchor a stiff silken stalk, perhaps half an inch in length, and attaches an egg onto the free end, like a punching bag mounted on a floor stand. Every hatchling drops into the world separately, with no chance to make a meal from its siblings. The mother's actions effectively preclude cannibalism.

THE RIGHT FOOD
FOR THE YOUNG

Outdoor observations convince us that insects possess an inherited guidance system. Usually, a flying insect is at the mercy of the elements. Just occasionally, a naturalist chances to arrive where visibility is good, the weather temporarily steady and propitious, and a flying insect takes advantage of the situation. One mid-afternoon in late July, on an alpine meadow well above treeline in Rocky Mountain National Park, Colorado, the wind ceased for half an hour. The sun from a cloudless sky warmed plants and animals alike. Gentle

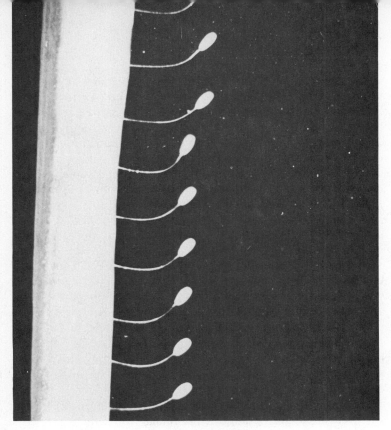

Each lacewing fly egg is supported separately on a stalk as slender as a hair. This prevents any hatchling from making meals of unhatched brothers and sisters. Each hatchling drops free and is instantly on its own.

heat from the thin soil began to stratify the air close to the ground, making distant vistas shimmer. Flies and bees buzzed about, visiting the flowers on some cushion-shaped alpine plants.

Suddenly five or six sheep-moths appeared, each beating its tawny wings vigorously. Rising less than six feet above the ground, these day-active moths traveled swiftly, one direction or the other, along a contour line. An individual might speed to our right for more than a hundred yards, then turn up or down for perhaps twenty, before whisking to the left for an equal or greater distance. We could not doubt that this was a

search pattern, and an incredibly systematic one at that. At
long intervals, a moth alighted. Within a minute or less, it was
airborne again. At last, one settled on a plant close to us. We
hastened to examine the actual target. Off went the moth, but
we hurried to the particular leaf on which it had clung. There
the insect had left a glistening egg. Through a magnifier, we
admired the bullet-nose form, vertical fluting, and creamy
opacity. We left the egg to hatch, confident that the mother
moth chose the plant with care by scent, perhaps then by touch
and taste. The weather let her find the food that her hatchling
caterpillars might devour and digest, toward perpetuating her
genetic heritage.

A majority of female insects have this special capability,
based on cues their sense organs pick up, to lay their eggs
where their young will find food, even though parent and
young may differ totally in nutritional needs. Generally it is the
parent of the caterpillar, not the caterpillar itself, that selects
the food plants upon which it will grow to full size. Perhaps this
explains how the larva can get along with only forty-eight
receptors for odors from its world, while the parent may have
thousands.

Larger moths of the same family (Saturniidae), to which the
sheep-moths belong, demonstrate clear preferences for specific plants in the American Northeast. The adults themselves
do not seem to feed, but flutter gently to street lamps and
lighted windows often enough to be known to many people by
name: cecropia, io, luna, polyphemus, promethia, and, in New
York City, the cynthia moth. Some of these have a greater wing
area than any other insect in this corner of the continent. Most
are softly colorful. Their big caterpillars, however, live mostly
high in trees, unnoticed until autumn, when they descend to
spin cocoons and spend the winter transforming into adults.

Saturniid caterpillars seem able to thrive on almost any kind
of hardwood tree, or even on an artificial diet containing no
leaf material at all. Despite this wide tolerance by the young,
a mother moth biases their selection. A female cecropia hunts
for willow, maple, or shrubs of the rose family. A female io
concentrates her progeny on wild cherry trees. A pale green
luna flutters to walnut, hickory, pecan, or persimmon if she can
find them. A polyphemus goes to oak, elm, basswood, birch,

or beech, but can be distracted easily to a sycamore tree. Promethia moths show a preference for shrubs of the laurel family, spicebush, tulip tree—or wild cherry if these others are all scarce. Only cynthia appears restricted—to the Asiatic ailanthus: "the tree that grows in Brooklyn." The tree and the moth were introduced separately to America from the same part of the Orient.

Chemical signals in appropriate sequence call forth a succession of actions among the female moths. First, the chemical ambience in her vicinity must be right to induce the female insect to liberate her own alluring fragrance, summoning a male of her species to mate with her. His scent promotes her cooperation. After mating, she customarily dismisses her partner. On a tree with the correct odor, she deposits a few eggs, then seeks out another tree with chemical appeal. Ordinarily too few moths of any one kind come to a particular tree for their caterpillars to menace the vegetation. Yet the behavior of these different moth mothers-to-be reduces the chance that the caterpillars of any two species among the related moths will compete for food. Over the years, this must increase the likelihood that each insect will survive.

The more we learn of the detailed behavior of insect mothers-to-be, the more examples we find in the almost endless array of adaptive features that qualify as peculiar, if not bizarre. One instance discovered recently pertains to the moths that are found among the long hairs of the sloth, a slow-moving mammal of tropical America. Many moths hide among sloth fur, but whether these adults gain any nourishment from the sloth directly, or from the algae growing in the hair grooves, has never been demonstrated.

A hunt for the caterpillars of this moth was undertaken by Jeffrey K. Waage and G. Gene Montgomery, using research facilities of the Smithsonian Institution on Barro Colorado Island, close to the shipping lanes of the Panama Canal. Many of the moths hid among the dense fur of the three-toed sloth, but not a caterpillar could be found. Closer study revealed that pregnant female moths of this kind would lay eggs in confinement, and from the eggs would hatch sloth-moth caterpillars. The larvae, however, would neither feed on sloth hair, nor on leaves of the three different trees that the sloths forage upon most frequently. The caterpillars would, however, feed avidly on sloth dung.

Further investigation revealed that the leisurely sloths descend routinely to the ground about once a week to void dung there—not in the forest canopy. The odor of fresh droppings attracts female moths from the sloth's fur, who lay their eggs on the droppings. If the sloth is still nearby, the moth flies back to its fur. Otherwise the insect wings high into the rain forest and searches for the same sloth or another.

The moth eggs hatch into caterpillars that commence to feed on this strange food. They ordinarily need several weeks to attain full size, to pupate in the remnants of the dung pile, before emerging as adults. These fly up to where sloths might be, find one if they can, mate among its fur, and continue the life sequence. Apparently the moths rely upon the sloth's hair covering chiefly as a shelter, and a place to find a mate, then upon the sanitary habits of the sloth to bring them to sites of fresh dung, among which their caterpillars can develop.

The solid wastes from larger animals attract insects of many different kinds in all parts of the terrestrial world where large mammals are native and the air is not too dry. In arid lands, the wastes dry quickly to dust and blow away, doing the soil little good, unless well-adapted beetles live there and benefit from transporting the dung underground. Appreciation for this role is spreading, largely because of educational campaigns where introduced dung beetles are being given a chance to demonstrate their inherited behavior.

A few years ago, Jamaicans expressed enthusiasm to us about dung beetles: "At last we've introduced something that does what it's supposed to, helps us, and harms nothing else!" Like other islands in the Antilles, Jamaica did not have large native animals to produce dung until cattle and horses were brought there a few centuries ago. Their wastes supported the growth of quick-breeding flies, but otherwise contributed almost nothing, unless people took the trouble to spread and plow under the manure. Now the grass-eating mammals and the liberated dung beetles will recycle the nutrients to the roots of plants.

The Jamaican farmers did not have much idea why the beetles should go to so much trouble. But then, fifty centuries ago,

The sacred scarab of Mediterranean countries shapes a mass of dung into a smooth ball and moves it to a place where the soil is soft enough to hide the ball underground and where the dung will nourish one beetle larva and the roots of plants.

even the most learned Egyptians scarcely understood the black scarabs they revered. All the way around the Mediterranean and up the narrow floodlands of the Nile, children saw these one-inch insects plodding along, propelling a ball of camel dung, supposedly from dawn until dark. To them, the sacred scarab mirrored the way in which the great god Ra rolled the sun across the sky each day. The head of the scarab, with which it shapes the ball, projects in points like the dazzling rays from the disc of the sun, and each of the six legs has five parts, making thirty joints altogether, like the number of days in a month. Yet no one seems then to have explained correctly why the beetle worked so hard with a ball of dung.

But ages passed before people realized that an animal might find nourishment in excrement. By burying a one-inch ball of dung away from the hot sun and drying air, a mother scarab safeguards a valuable resource for the single fertilized egg she deposits on it. Her offspring will have no need to hunt for its meals and will grow in solitude and emerge into air as an adult beetle. Much of the dung will actually be left over, as fertilizer where the roots of plants can use it best.

A preference for dung beetles over hordes of noxious flies has led recently to international cooperation. South Africans have built the Dung Beetle Research Unit in their country with Australian funds. Its field officer, G. F. Bornenissza, collects various insects he hopes will settle down and breed in captivity, so that eventually large numbers of dung beetles may be raised. The ultimate aim is to introduce suitable insects into arid parts of Australia to bury the droppings of cattle, sheep, and horses.

The great continent "down under" does have a dung beetle of its own, but it restricts itself to marsupials, and its grubs get wastes before they are voided, by clinging to the lining of the rectum until they are ready to pupate. The development inside the mammal does no harm, but works no wastes into the ground either, to benefit the plants.

CARRION AS A RESOURCE

Among our treasures from South America is a cobalt-blue scarab from Brazil, an inch across, almost as much in depth, and more than an inch and a half in length, counting the short horn upon its head. When alive the insect weighed about as much as a house mouse. It came reluctantly from a vertical chamber, a cylindrical shaft it had excavated below the inert body of a dead python. Laboriously the female beetle had dug her hole. She had carried to its bottom and consolidated about a handful of snake meat. She would have laid her egg on this food supply and filled the hole with earth, had she not been interrupted. Her single larva at the site, buried alive but out of the way of competitors for the meat, could have attained full-fed adulthood in this particular inherited way of life.

Much smaller and more widespread beetles attend to the disappearance of small carcasses, such as those of mice and rats, birds smaller than a robin, and snakes of modest length. Seldom does anyone give them credit for their labors, which add much nitrogenous fertilizer to the soil, or even wonder what happens to all the familiar small creatures that die of accident, exposure, or disease. The beetles that inter small carcasses are unique in extending maternal care so far and in having the aid of the male so often until the larvae are ready to pupate. Known as burying beetles, or as sexton beetles from times when the duties of the sexton at the church included grave-digging, these insects seem extraordinary in the plasticity of their behavior and in the success that it has brought them.

The faintest odor from a small corpse, perhaps emanating before the body of a bird or mammal has cooled after being struck by an automobile, brings the beetles flying. The first arrives, often in less than half an hour, examines the trophy, and sets to work to bury it, either where it lies or nearby in softer soil. Haste and hard labor are essential. A member of the opposite sex may alight a short distance off, and rush over to participate, but the beetles take no time to court, to mate, or to feed until the carcass is secure underground. The location of the corpse is usually well concealed, and the soil itself absorbs any odors that might reveal its location.

We have watched burying beetles for many hours, and put them to various tests. We can confirm completely the careful observations made almost a century ago by the French naturalist Jean Henri Fabre. He admired these little gravediggers of the animal world, describing them as being "elegantly attired" in black, with a "double, scalloped scarf of vermilion" across their shining wing covers. Yet an observer cannot readily watch this handsome beetle for long. A burying beetle slides quickly out of sight below the carcass it has found. Then, lying on its back, the insect uses all six of its powerful legs as levers to shift its prize. Or if the conditions are suitable, it propels itself with all its legs at once to bulldoze headfirst into the earth, to loosen the soil and to push the diggings from under the carrion. Inconspicuously, a fraction of an inch at a time, the carcass moves horizontally or sinks into the ground.

The carrying of small corpses from hard ground to soft, in one steady direction, a little at each move, impresses anyone because the insect is so small—usually less than an inch in length—in comparison to its load. The spectacular activity led the Danish entomologist J. C. Fabricius to apply the generic name *Nicrophorus* to the insects in 1781, translating the Greek roots *nekros* and *phoros* as "corpse bearer."

Some measure of the success gained through this behavior can be seen in the fact that the genus *Nicrophorus* includes almost one hundred species, with some overlap in distribution. About half are Asiatic, from the arctic USSR to the Near East, Sri Lanka, Thailand, China, Japan, the Philippines, the Solomon Islands, New Guinea, Celebes, Java, and Sumatra. Fewer species inhabit Europe and North Africa. The New World has *Nicrophorus* from Alaska to Newfoundland, and southward all the way to Tierra del Fuego. So far, none has been found in Australia, New Zealand, or the West Indies. The fossil records extend their history back to the mid-Pliocene period in northern Germany. Several kinds have been recovered from the semifluid asphalt of the famous La Brea and McKittrick tar pits in California, along with the remains of ancestral coyotes, sabre-toothed cats, dire wolves, giant ground sloths, and other extinct beasts. A date of mid-Pleistocene has been assigned to these finds—a few thousand years before the earliest human hunters arrived on the scene.

All burying beetles become adult in the size range from $\frac{3}{8}$ inch to $1\frac{3}{8}$ inches, each species showing a wide variation in size. Yet the smallest of these insects seems as ready as the largest to transport the compact body of a bird or a mammal weighing up to $3\frac{1}{2}$ ounces, such as a robin or a rat. Anything heavier is usually abandoned, unless it is only slightly overweight and can be interred on the site. The corpse of a marmot or a fox is too big. A snake three feet long, however, can weigh more and still be buried expeditiously. Its body is subdivided into two or more zones of operation. A pair of *Nicrophorus* beetles attends to each zone. Often the tail is left sticking out of the ground. By noticing the neglected, projecting tail, we have discovered buried snakes, tailed amphibians, even slender fishes such as eels that fishermen have discarded. Neither ants nor blowflies pay much attention to an appendage as unmeaty as a tail. But by opening the earth with a trowel, it is easy to find the burying beetles where they are still in attendance.

Nicrophorus beetles are by no means the only insects that sequester nonliving food for their larvae, but they usually work as a team, whereas the others (the sacred scarab, its relatives, and various solitary bees and wasps) work alone. Either a male or a female *Nicrophorus* will initiate the flexible behavior that gets the larval food into a safe place. Often, when the firstcomer is small and the prize it finds is large, it will cease operations for as long as thirty minutes to stand vertically, head down, while secreting a volatile lure from the posterior end. We have seen as many as two dozen more beetles arrive from downwind, all of them potential mates—and all stirred to combat the others until one victor remained to complete the pair.

A partner is accepted with no time off. The two labor together at intervals, and also separately in loose cooperation. Either member of the pair may creep into a more or less concealed place and appear to sleep for as much as half an hour, or depart on feet or wings to some unknown destination for a comparable period, only to return and resume work. Ordinarily, copulation is deferred until the beetles are securely in a chamber of their own making, an inch or more below the surface of the ground.

Other male insects would perform their brief sexual duty and depart; the inseminated female would carry on alone to the end of the sequence of behavior specified by inheritance. Occasionally burying beetles follow this pattern, but usually both parents remain. Together they work on the mass of food, freeing it of fur or feathers, and perhaps adding secretions that modify the course of decomposition. Gradually they push the corpse into an almost spherical form.

As the insects clamber around the carcass, which supplies food for them as well as for their young, the walls and roof of the earthen chamber become firmly packed. The female constructs a short vertical extension of the chamber above the carrion and lays her eggs in the side walls of the passageway. She returns to the carcass, and by a combination of selective feeding and clawing at the upper surface prepares a conical depression. Both beetles regurgitate into the depression droplets of partly digested tissue. The fluid accumulates as a pabulum for the larvae that soon will hatch.

Fabre, or any other patient observer, could discover this much by exhuming the beetles and their food supply at the proper time, just before the young hatch. But Erna Pukowski, studying species native to her Polish countryside, managed to make captive burying beetles so much at home, notwithstanding the unnatural conditions she created by illuminating their burial chamber, that she could follow the next steps.

One beetle (perhaps the female, although the members of a pair are too much alike externally for an observer to distinguish sex) stood beside the pool of liquid nourishment and began to stridulate. The sound brought hatchling larvae, each some two or three millimeters long and almost like maggots in appearance, to the parent's side. The beetle sipped from the pool and then transferred the fluid food to one larva after another. The larvae lifted their mouth ends, the better to receive the food. Sometimes both parents shared in the feeding operation.

In 1972 Carsten Niemitz of Justus Liebig University in West Germany discovered that very young larvae will orient themselves to the sound of an adult's stridulation recorded on tape. This response disappears, however, after the larvae have molted for the first time. Even so, older larvae renew their solicitation of regurgitated food for a few hours after each molt by approaching any adult that is close to the pool of food and pressing their mouthparts against its jaws or palps. This action stimulates regurgitation as before. Otherwise the growing larvae feed directly from the pool or pull fragments from the wet surface of the carrion.

The larvae receive parental care all through their period of feeding and growth. The parents may even prepare a horizontal passageway into which the fully developed larvae crawl to pupate. Only then, when the adults can contribute no more to their brood, do they force their way upward through the soil and fly away.

We have not yet marked and followed the departing parents to learn whether they repeat the same sequence of behavior elsewhere. They probably do, since adult beetles live from three to fifteen months, depending on the species. They search widely for the odor of recent death and are remarkably efficient. Frantisek Petruska, a Czechoslovak ethologist, has found that by marking beetles captured at carrion bait and releasing them at various distances, some will return to the carrion within twenty-four hours from as much as four kilometers away.

Competition for small carcasses is frequently intense. Ants and flies tend to take over during the day. Burying beetles of the species that are most active by day succeed only if they can inter a body quickly. At night, the competition is mainly from other *Nicrophorus.* The largest beetle present generally repels all the others except a mate. This is probably why we find two beetles of the same size cooperating much less often than a large male with a small female, or vice versa.

Burying beetles have other ways to reduce competition. Each species has a preferred combination of temperature range and relative humidity. This pattern, as Jean Théodorides of the University of Paris showed in his laboratory, keeps certain beetles in woodlands and others in open fields. Burying beetles that are active in the spring belong to species that go through the winter as adults, whereas the beetles found competing in the summer are likely to represent species that spend the cold months dormant as pupae or full-grown larvae.

Animals that eat carrion and insects too constitute a special hazard for burying beetles. In one Ontario woodland, striped

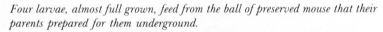

Four larvae, almost full grown, feed from the ball of preserved mouse that their parents prepared for them underground.

skunks proved incredibly efficient in finding each mouse carcass we left out overnight. Raccoons that forage along the
shores of lakes and rivers and the Atlantic coast clean up fish
so quickly that this food is no longer available to burying
beetles. This is blamed for depriving North America's largest
burying beetle of an essential resource until the species has
become rare if not extinct.

Noting the number of different obstacles a typical environment is likely to present to beetles as they try to move or bury
a body, Fabre wrote that the insect therefore "cannot employ
fixed methods in performing its task. Exposed to fortuitous
hazards, it must be able to modify its tactics within the limits
of its modest discernment. To saw, to break, to disentangle,
to lift, to shake, to displace—these are so many means that are
indispensable to the gravedigger in a predicament. Deprived
of these resources, reduced to uniformity of procedure, the
insect would be incapable of pursuing its calling."

Once we drove a good-sized stake into the ground at a 45°
angle and attached a strong cotton string to its upper end. We
tied the dangling end of the string around a hind leg of a dead
mouse lying on soft ground. A pair of *Nicrophorus* beetles
pushed away the soil below the body until the mouse hung

*A burying beetle gnaws on the string tethering the hind leg of a dead mouse,
having discovered what was preventing the body from descending into the hole
dug from beneath.*

from the tethered leg over a cup-shaped depression. The insects cleared a space the thickness of their bodies between the mouse and the soil and then kept swiveling the carcass in wide arcs. The tail of the mouse dragged on the rim of the depression until one of the beetles chewed it off.

That did not solve their problem, and both beetles continued to explore the surface of the carcass. Only about six hours after they began work did one of them discover the tether. In less than a minute the insect settled down to gnaw through the cotton fibers. By dawn the carcass had been liberated and buried.

To test the strength of *Nicrophorus* beetles we rested one end of a flat rock on the body of a two-ounce (fifty-gram) mole. The rock applied more than a pound (about half a kilogram) of unyielding weight to the body. Two beetles were nonetheless able to work the body free. First they took up positions side by side with their backs against the rock and their legs against the body. They shifted the body about three-eighths of an inch (a centimeter) in relation to the rock and then repeated the performance with respect to the hard soil below the body. Alternating between these two areas of contact, they freed the carcass in less than half an hour, whereupon they transported it to soft ground and quickly buried it.

In tests of the memory of burying beetles we have found that if a beetle has had fifteen or twenty minutes of experience with a suitable carcass, it can be removed and held captive for at least sixteen hours without losing its readiness to return to the body within minutes of being released. After twenty-four hours of separation from its trophy the beetle is more likely to fly off.

The patterns of behavior inherited by burying beetles reveal no sharp demarcation between solitary and social activities. The solitary mode seems dominant through much of the insect's adult life. The beetle comes and goes where olfactory cues alert it to decay, in fungi, dung, and carrion, including carcasses that are too large to manage. Each *Nicrophorus* feeds independently, showing no obvious reactions to mature members of the same or other insect species unless it is working to sequester food for young. Upon occasion, they attack and devour fly maggots. In Europe, burying beetles have been praised by potato farmers who saw adult *Nicrophorus* beetles killing and feasting upon the larvae of the introduced Colorado potato beetle.

One characteristic of burying beetles tends to distress anyone who admires these hardworking insects. The adults are "abominably verminous," even "in the company of the larval family," as Fabre pointed out. The vermin are small red mites *(Poecilochirus)*, which swarm over the body of each beetle, although rarely on its antennae. An English parasitologist, B. P. Springett of the National Environmental Research Council in Alston, seems to have explained the presence of the mites as part of an amazing mutual-benefit association. Entomologists had assumed that the mites nourished themselves by inserting slender mouthparts between the body segments of the beetle to obtain a blood meal. Instead, the mites appear to sustain their growth and reproduction at the expense of fly eggs and minute maggots which otherwise might ruin the food reserve for larval *Nicrophorus*. Certainly the mites rush from a burying beetle that finds a dead mouse, and run over or through the fur while the body is being transported or buried. Perhaps the mites also sip from the pool of regurgitated nourishment the beetles provide for their young. The cost to the beetles would be small by comparison with the damage that fly maggots could do. Ordinarily, a large proportion of the food mass remains after the beetles have raised their brood, as a contribution to soil fertility.

A multitude of mites accompany each *Nicrophorus* larva as it gets ready to pupate. They wait within its pupal cell, then clamber all over the adult beetle as it emerges into the open. No fewer than 405 mites clung to the body of one burying beetle examined near Cracow, Poland, according to a count published by Jerzy R. Starzyk in 1967. The number strained our credulity so much that we picked up three particularly "verminous" adults of *N. marginatus* in New Hampshire, transferring them quickly with metal forceps from the ground into a vial of alcohol preservative. Our forceps dislodged a small number of mites—perhaps thirty-five—or they tumbled off as the beetles struggled for the second or so before falling into the vial. We made no attempt to catch the mites that dropped, for they burrowed into the soil. But we did find 1,236 mites that held onto the three beetles, pick-a-back or pick-a-belly, all the way to the preservative, an average of 412 per beetle. Even if, as Starzyk estimated, 95 percent of the mites fall off during a five-minute flight by the beetle, more than a dozen would arrive to assist with fly eggs and maggots.

The social habits of burying beetles fit between the extremes shown by other insects. The pattern followed by the parent or parents that attend only to their own offspring seems most primitive. Among more advanced social insects, a female, at least, provides care for the offspring of other females, often as a sterile surrogate parent. Burying beetles often show some altruistic behavior in that small members of the same or different species may contribute importantly to the rapid burial of a carcass, then depart with no share in reproduction. The dominant mated pair takes over the entire food supply. But they tend only their own larvae, not those of a different female.

PARENTS WITH A LITTLE EXTRA

The combinations of adaptive features in behavior and body build among insects provide many ways to improve the opportunities for young and hence the chances for survival of the species. Sometimes one parent or both carry a contribution toward the food supply with them to the site where eggs will be laid and larval growth accomplished. The practice reminds us of prospectors in the West who took along a small container of sourdough starter to leaven future meals of bread and biscuits. The insect brings a different fungal partner, one that will make plant products more digestible.

The most conspicuous of insects with this behavior in our part of the world is a horntail wasp. The adult female, which is almost two inches long, flits about in search of dying and dead trees that are still moist inside. Carefully she thrusts a needlelike ovipositor into the wet wood and deposits eggs that will hatch in a few days as grublike larvae. Along with each horntail egg go a few cells of fungus from tiny pouches at the base of the mother's ovipositor.

The fungus needs an hour or less to respond to its new location, attacking the wood, extending mycelial threads into it, and digesting the hard material. Upon hatching, the larva of the horntail wasp feeds on both the fungus and the infected wood, tunneling wherever the fungus is active, until it attains full size. It then constructs an egg-shaped cocoon of wood chips just where its tunnel, the diameter of a ballpoint pen, opens to the outside world. An adult later emerges through this escape hatch. If a female, she will carry a supply of the fungus and be ready, after mating, to repeat the cycle. The larval female possesses pits between the first and second segment of the abdomen. A waxy secretion within these special pits traps fragments of fungus and keeps them alive, even if the fungus in the wood of the tree dies from drying out. Subsequently, when the horntail larva transforms to a pupa, the pockets of wax and live fungus break open. Flakes of the waxy material, still impregnated with fungus, become lodged in the moist pouch at the base of the immature female's ovipositor. There the fungus reacts, forming the special infective bodies that may later stick to any eggs she lays.

A totally different insect that relies on fungus often riddles hardwood trees with small openings, called "shot holes" from their size. It favors trees felled by a hurricane or weakened by drought, disease, or the ravages of foliage-eating animals. The holes mark burrows of ambrosia beetles, each scarcely a quarter-inch in length. Large numbers of these insects often fly upwind in the evening toward any tree that has been severely cut or bruised. They detect the odor of fermenting sap diffusing into the breeze. Each beetle alights, perhaps to use its hard jaws to cut through the bark. The insect does not eat bark or the underlying wood, however, but discards a steady stream of tree dust, mixed with fecal pellets.

As the beetle labors to cut inward, many live spores of the important fungus drop from pockets near the base of the first pair of legs. The fungus germinates as the insect deepens its cavity and clears out the debris. The fungal strands spread rapidly in the tunnel. In a day or less they coat the inner surface with a velvety pile. The protruding cells are the only food these beetles or their young are known to eat. To them it is ambrosia.

So long as the ambrosia beetles or their larvae are present, releasing secretions and excretions, the fungal cells appear almost yeastlike in their spherical form. Without this chemical influence from the insects, the fungal cells extend as slender cylinders. Perhaps the secretions and excretions also serve to prevent the growth of competing fungi for, among all the ambrosia beetles that have been carefully observed so far, only one species of fungus associates with each species of beetle.

The male in some species of ambrosia beetles stands aside while his mate excavates the tunnel. In others, the male assists her not only in extending the main passageway but also in cutting a series of short side branches beneath the bark. She lays one egg in each side chamber. The larva that hatches there eats only its local supply of fungus. Still different species encourage togetherness. The parents excavate a few large chambers, within which several shining larvae feed on the fungus, still accompanied attentively by the mated pair.

No one seems to know whether woodpeckers and other birds that seek insects on or below the bark ever notice the distinctive odor produced by the beetles and their fungus. The sense of smell probably is too poorly developed in these particular feathered fliers. The shot holes, however, where bark beetles enter and escape, would be much more likely to attract the birds' attention, for their sharp eyes are perfect for detecting such fine detail. Yet some trees in which hundreds of bark beetles are busy feeding show scarcely a shot hole at the surface. The diminutive male parents stand, like living corks, stoppering their tunnels through the bark. Their bodies have a truncated rear end, which they keep flush with the surface of the tree all day. They conceal the presence of the tunnel, which branches extensively within the innermost and highly nutritious tissues of the bark. There the females and young are feeding.

We had to go to central and southern Africa to meet the insect mothers that give their young the most incredible care of all. They are the females of the dreaded tsetse fly, a bloodsucker that transmits the infective agents that cause sleeping sickness in people and the disease known as nagana in domesticated livestock. The sharp-eyed fly rests all night and much of every day in the shade of some leaf, watching for some moving vertebrate animal from which it can get blood. The host may be an antelope, a giraffe, a zebra, a fish eagle, or a crocodile. These and other native animals are immune to the disease the fly carries. Yet they serve as natural reservoirs for the infective agents. The khaki-colored insect itself is deceptively quiet, less than half an inch in length. Its bite is almost painless, and the amount of blood it takes is small. The tsetse is so unwary that it can be killed merely by pressing a finger on its back!

Efforts to open large areas of Africa to cultivation and ranching activities through eradication of tsetse flies led to discovery of the insects' peculiar mode of motherhood. Both sexes of the fly come in modest numbers to large grazers, attracted by movement, then by smell. The insects settle to feed from shady underparts of their host, and there they also meet a mate. Their union lasts an hour or so, and supplies the female with all the sperm she will ever need.

Tsetse eggs develop one at a time and hatch inside their mother. The solitary maggot lies in a broad part of her oviduct, and takes nourishment from her through a sort of teat. Progressively, her one larva grows to full size, until it is actually heavier than the mother. What appears to be birth is merely the escape of the larva from its parent. It drops to the ground from the bush or tree on which she happens to be resting, and within fifteen minutes, it has burrowed into the soil. The larva's outer skin hardens, preventing desiccation. Transformation to the adult body proceeds rapidly. In two weeks or less, the mature fly bursts from its larval skin (puparium) and forces its way upward to freedom. Meanwhile, another maggot develops inside the mother tsetse. She may produce eight altogether during her normal three-month lifespan.

Tsetse flies and their young cannot tolerate the heat from the full sun or the temperatures of summer in equatorial parts of their range. The maggots ride along, individually concealed, as their mothers fly seemingly at random but actually more in a southward direction in November. A few months later, other larvae in subsequent females will be carried progressively northward. The individual trips are brief—no more than two minutes' flying time from one stop to the next. The mother needs an hour's stopover while she digests more blood and refuels her huge wing muscles for the next move. Yet her bias in direction permits migration. It extends the heritage of tsetse flies into fresh areas, as well as into a continuing future.

MIGRATION

The repetitive behavior of many insects in coming home, perhaps to the same place each night or the same area each year, resembles human ways. Yet the insect has built-in advantages. The compound eyes of some insects can detect, as those of vertebrates cannot, bright and dark patterns in blue sky that provide the creature with the equivalent of a compass. We need a Polaroid filter to see this pattern. The sky compass gives guidance to insects with a homing habit and others that follow a migratory pattern. These travelers also possess an inner (and still unexplained) sense of time; it lets them correct their heading according to the time of day. The insect automatically allows for the rotation of the earth, which causes the sun's position to course from east to west across the sky, turning the plane of maximum polarization at the same rate.

No normal ant, wasp, or bee, it seems, emerges from darkness underground or in the nest without looking carefully at the sky, perhaps correcting the inner clock and taking note of a few landmarks. Its memory holds this information and serves also as a flight recorder, even while the creature is afoot. If we imprison the traveler for an hour or two in a light-tight box, the insect may struggle to escape or quiet down, but in either condition, its inner clock continues to move at the regular pace. Upon being released, the insect needs only to see a patch of blue sky to know the most direct route home. An overcast sky offers no guidance in strange territory, and a bee that is freed from a closed box there merely waits—until the next day if necessary. It rests quietly until the clouds part and the sun makes the polarization pattern evident again.

Unrelated insects navigate with a sky compass too. Ladybird beetles hunt time after time along the same stalks and leaves searching for plant lice and other small prey, as long as the sky is overcast, but fly to new territory as soon as clear areas of blue appear. The common drone fly, which so closely resembles a male honeybee, also directs its course among the flowers and foliage of a favorite region with cues from unclouded sky.

We often mark an individual fly with a harmless spot of colored paint, and see the identical insect on successive days for weeks —probably for a major part of its lifespan as an adult.

With similar guidance, individual heliconiid butterflies in the American tropics (such as the zebra, black with yellow stripes) direct their flittering courses each evening to the same chosen branch. There they sleep through the night in close proximity, like children at a pajama party. Every evening we can watch the identical insects arrive and seek their particular places among dozens of others. They remain almost motionless until long after the sun is up, then burst into flight as though an alarm clock had gone off. By midday they are all hundreds of yards apart.

Using the sky, some dragonflies, moths, and butterflies migrate regularly between a summer range and winter quarters. Their heritage in each generation calls for a general heading, one for autumn, and an opposite one through spring and summer.

The green darner, a common dragonfly with a four-inch wingspan, follows a route roughly northeast and southwest between Florida or the Gulf Coast states and New England or eastern Canada. The same dragonfly may make the trip south and north again. Despite its tattered wings, it finds a mate and starts a new generation in ponds fed by melt-water from winter snow. The migratory moths and butterflies, by contrast, appear normally to have one southbound generation that reproduces in the warm climate, and another (or more) that disperses the species north again.

The first butterfly to have its travels traced is still the most wide-ranging. Known as the "painted lady," the adult occasionally overwinters as far north as southern England. It deposits eggs on every continent except South America. Only in the Old World, however, is this insect a regular migrant. A winter brood develops between December and May on wild vegetation across North Africa to northwestern Pakistan. The adults fly northwestward through Iraq or across the Mediterranean Sea, and through the high passes of the Alps like a chain of flowers. By the end of May they are common in much of southern England and Wales. Occasional individuals fly beyond the Arctic Circle in Scandinavia. Summers of special abundance, such as 1966 in Britain, are recorded as good years, in the general category of short-lived phenomena. They are usually followed in autumn by reports that vast numbers of these butterflies are traveling south, aiming for the same destinations their ancestors found.

C. B. Williams, a British entomologist who pioneered in the study of migratory insects, says that only one person has recorded the beginning of a northbound flight of painted ladies. The date was in March 1869, and the place the desert behind Suakin on the Sudanese coast of the Red Sea. A Mr. Skertchley noticed a sudden agitation in the tall grasses through which he was riding on his camel. He dismounted to learn what could move the vegetation on such a windless day, and discovered "myriads" of painted ladies emerging from their chrysalises and drying their wings. Fascinated, he watched for half an hour as most of the insects flew off eastward to the Red Sea coast.

At least a million moths winged south through the Kaiser Tauern pass of the Austrian Alps on the night of July 24, 1966. A sampling of the mass travelers, attracted in just two hours to an illuminated trap, revealed 17,290 of one kind of cutworm moth, 1,520 of another, 76 of a third, along with 114 assorted other insects that probably came from the immediate vicinity. All three kinds of cutworms are pests of field and garden crops. No one could imagine how so many of them congregated on a single night to fly together up one small valley and over the pass at an elevation of 8,200 feet. A closer look showed that the females were all young adults in fine physical condition. Maturation of the eggs in their ovaries was just beginning.

The commonest of these prolific insects travels by day as well. Hordes of them reach North Africa, where they mate and lay their eggs. Later, the winter rains promote the growth of vegetation along the fringe of the great deserts, and the caterpillars can emerge and feast. Each stores nourishment in its body as a reliable, transportable reserve for northward travel after the creature transforms to the moth stage. Not until mid-May is the weather likely to stimulate the adults to fly, but then they cross the Mediterranean Sea and reach Europe. Reproduction can take place in whatever country and at whatever latitude the individual moths select.

The cutworm moths and their caterpillars earn the ire of agriculturalists in Europe when the larvae feed on peas, beans, flax, and corn, although the same species also attack thistles, nettles, and other weeds, plants that were probably their natural foods long before humankind invented agriculture. Many of the caterpillars reach full size on any of these varied diets. They spin little webs just above the soil, and transform in these flimsy cocoons, to the gray-brown moths. By the end of July they are ready to start south. The two-way migration, according to season, involves two generations and quite different food plants in each round trip.

Occasionally some migrant species of attractive insects arrives in spectacular numbers to cluster harmlessly year after year. In some instances, they can become a tourist attraction. The town of Petaloudes on the Greek island of Rhodes advertises its "Valley of the Butterflies," and a community along the California coast urges "Follow the butterflies to Pacific Grove."

The "butterflies" of Petaloudes prove to be tiger moths with a wingspan greater than two inches. Resting by the thousand on bushes and bare embankments, the insects create a restless mosaic. No two moths wear identical patterns. Sometimes even the yellowish-white streaks on the dark brown body and forewings differ slightly from right to left. When a moth stretches its forewings to reveal the hind pair, they flare brightly red with bold black spots. These moths of Petaloudes arrive individually at night in June and July each year. By August, not an insect of this kind is to be found elsewhere on the island. Yet in September they depart again, once more in darkness, each female freshly mated and on her way to some distant weed on which she will lay her ivory-colored eggs.

Rhodes is a little island, only 540 square miles (half the size of the state of Rhode Island), the easternmost of these areas surrounded by Mediterranean waters, and separated by only ten miles from the coast of Turkey. Moths of the same species are found on other islands and on the mainland from Asia Minor and Egypt to Belgium. Yet only on Rhodes have the travels of the insects been traced and their clustering in a cool valley during the hottest of the summer become a cause for celebration.

The tiger moths of Rhodes and the monarch butterflies of Pacific Grove are all adults with an inner reserve of stored food left over from their caterpillar days. Reproduction is still to come. The insects may flutter occasionally from the throng to replenish their moisture with a sip of dew or nectar, but thirst, rather than hunger, impels the individual toward this brief activity. The inherited schedule of life for the moths calls for resting through months when even the weeds shrivel in the heat. The program of the butterflies at their California clustering sites puts their lives on hold while winter farther north ends all growth of the milkweeds on which their caterpillars munch. The monarchs come mainly from northern parts of the state and western Oregon, and settle year after year where insects of their kind have found freedom from repeated frosts. They roost, like a covering of strange flowers, on tall evergreen trees close to the shore. Some choose Bodega Bay. Others travel to San Diego, and a few beyond into Baja California. But only in Pacific Grove is the arrival of the butterflies welcomed late each September, their safety promoted by an ordinance authorizing a $500 fine for anyone caught interfering with the butterflies in any way. By April the insects are northbound, and the butterfly-roost trees seem dark and bare.

East of the Rocky Mountains across southernmost Canada and the northern states, the sturdy monarchs have long been noticed making mass flights. The adults congregate each evening on tall trees in late August and September, then take advantage of the next clear morning to stream south. Those that reach the north shore of Lake Ontario after midday rarely strike out across the open water. Instead, they perch on the oak trees and wait until the morrow. By sunset in September the oak branches are often bowed down by the weight of butterflies, each clinging to its support until warmed by the sun next day. In rain or storm the insects stay where they are and scarcely move. They need sun and blue sky for heat and guidance before crossing the lake.

How does an insect know how broad the water is, how long a butterfly must keep airborne to reach the southern shore? Do the same insects return, or is this a kind of tribal memory stored in the genetic heritage? These and many related questions challenged Frederick A. Urquhart of the University of

Toronto. He took advantage of his location and made studies of this migratory butterfly a lifelong enterprise. His wife Norah and, for a while, his son Douglas too, helped as he caught and tagged the insects, then tried to record their travels southbound across the United States. Gradually he perfected his techniques and enlarged the number of cooperators in distant places. People all over the continent began to watch for, capture, and report the information carried by tagged monarchs, without interfering more with the continued travels of the insects.

We joined Fred and Norah in the darkness before dawn one September morning and helped carry the big cardboard cartons while Fred led the way with his long-handled net. His tool let him reach up into the oak trees near the lake shore and catch the sleeping butterflies in the twilight. One netful after another was transferred into the cartons, to be hauled back to the screened porch at the Urquhart home. There we all got busy soon after the sun rose. Release a few dozen butterflies from the darkness of the boxes, and let them settle on the screens. Pick off one insect at a time, gently holding it by its white-spotted black body while using thumb and forefinger to softly rub off the scales from near the base of one forewing along the leading edge. Both top and bottom surfaces must be bare. Then a self-adhesive printed tag of waterproof white paper should be folded around the edge of the wing and pressed to the bare areas. Now flight-test the butterfly by freeing it to fly to the screen. Does the tag hinder symmetrical use of its wings, and need repositioning? If not, get another butterfly.

When finding an untagged monarch became time-consuming, we picked up half a dozen at a time in each hand, took them outside the door, and let them go. Steeper than the takeoff of a jet airplane, they flew upward and headed south —a line of swift-flapping butterflies departing Canada for states across the lake.

For a while, the record for travel was held by one monarch tagged on the north shore of Lake Ontario at Highland Creek on September 18, recovered on January 25 at San Luis Potosi, Mexico, about 1,870 miles to the southwest. Another butterfly tagged the same day in the identical place reached Roxie,

Mississippi, on October 5, having traveled at least 1,060 miles in seventeen days. Did the insect keep to a straight route—a "beeline"—or digress along the way? A third individual from that particular morning's catch turned up in Niceville, Florida, after twenty-three days, at a distance of 985 miles.

When the Urquharts and their collaborators tagged monarchs at overwintering sites in Florida and California, only the females among them were ever seen again farther north. Evidently the butterflies mate at the end of their protracted rest, but the males go nowhere. Fertile females fly part of the way north in early spring, find young milkweed plants already sprouting, lay eggs, and die. In due course, the caterpillars feed on the milkweeds, mature, transform, and emerge as colorful adults; they mate, and females of a new generation head north or northeast. By the time pregnant monarchs arrive in northern states, the later spring there has provided the insects with milkweed plants that are still young and succulent. A third generation may be needed to send females into Canada—as far as the shores of Hudson Bay. Only the brood that matures around the autumn equinox sends migrants of both sexes southbound. Perhaps the relative lengths of night and day hold significance for butterflies, as well as for birds and people. No one can tell how the insect will direct its flight without letting it free. And then catching it again is not so easy, if it sets off on its inner-directed journey.

Roosting areas in Florida and California gave sanctuary to only a minuscule fraction of the monarch butterflies that winged south. Where else did the millions wait out the early months of the year? In 1975, the Urquharts found part of the answer. So many of their collaborators reported migrating monarchs crossing the southern borders of the United States in huge numbers that an intensive search in Mexico seemed urgent. But where? The forested tops of mountains seemed more probable havens than any land at lower elevations. No roads led there. Even a farmer swinging his sharp machete might regard as a waste of time the effort to penetrate the underbrush in regions so inhospitable to human uses. Yet labor of this kind, subsidized from limited resources during a winter leave from teaching duties at the university, paid off. Monarchs by the millions awaited discovery where they clung

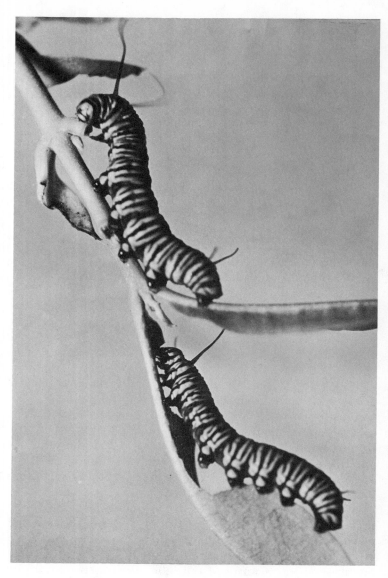

Monarch butterfly caterpillars nibble on a milkweed.

These chrysalids, plus caterpillars and a banded adult, were raised from larvae in captivity.

A freshly emerged monarch clings to its chrysalis.

A male monarch with wings outstretched is ready to fly a thousand miles or more. The black spots close to a linelike vein in each hind wing identify it as a male; they are pockets in which a tranquilizing lure is placed by the male to subdue a female until he can get into a position to mate with her.

to trees and shrubs near the crest of a remote mountain, a peak
in the Sierra Madre north of Mexico City. Among the throng
were butterflies with tags, including one made distinctive by
Jim Gilbert in Chaska, Minnesota.

Before leaving the mountain sanctuary, Fred and Norah
tagged thousands of the monarchs there in Mexico. A satisfy-
ing number were recovered the following April in northern
Texas, a thousand miles from the wintering site. Every one of
these travelers was a female, on her way to rendezvous with a
milkweed plant as far from Mexico as spring weather would let
her go to place her eggs.

One year passed without any revelation as to the exact wintering site of so many monarchs. The Urquharts told of their
discovery only in a deliberately general way, while encourag-
ing the Mexican government to declare the site off limits—a
national treasure. But a second team of scientists analyzed the
vague publication and gained enough guidance to hack a second narrow trail up the same mountain top. With more time
and eyes to study the phenomenon, the new visitors saw Mexican residents already exploiting the chilled butterflies. Ranch
hands used blocks of salt to tempt cattle up the narrow trail,
and shook the bushes to create a hail of insects, all chilled to
helplessness. The cattle ate the monarchs by the thousand.
Other mountains in the same country provide wintering sites
for these migratory butterflies. Yet timber-cutting operations
and the spread of people with domestic animals upslope
threatens to destroy these sanctuaries too. Some of the local
insect-eating birds have already learned to discriminate between monarchs that contain poisons and those that developed from caterpillars on milkweeds with lesser amounts of
these distasteful, dangerous substances.

The monarchs themselves seem unlikely to disappear, although their two-way migratory drive could well be lost as an
essential adjustment to a changing world. In many parts of the
earth, these particular butterflies already have dispensed with
mass migrations, while retaining their travel habits for emigration. Long ago, the monarchs found food plants suitable for
their caterpillars in the West Indies and South America. Most
of them stay at home there except in years of special abundance, when millions fly away. No one is sure how many of

them perish on their own, and whether some succeed by hitch-hiking aboard a ship to reach a new destination.

ONE-WAY TRAVELERS

The first monarch to be seen in Britain appeared there intact and active in 1876. Between that date and 1962, a total of 214 more were caught in England, as many as 30 in some years. The Isle of Wight, close to the port of Southampton, proved especially attractive. At first, naturalists calculated that a strong wind might keep the fliers high above the ocean and propel them eastward from Canada, perhaps at an average speed of fifty knots. At least fifty hours would be needed to cross the North Atlantic. Would a butterfly continue in darkness, or keep flying if lifted to moderate altitude and chilled? Might it settle for the night on a wind-tossed sea and take off again at dawn? A ride as a stowaway seemed far more plausible.

This same kind of insect is known as the "traveler" in the Pacific world, where it has turned up on remote islands from time to time. If a pregnant female could find a milkweed plant introduced from America, she started a resident population of the butterfly. Monarchs colonized Hawaii in 1845, the Marquesas in 1860, New Caledonia and the east coast of Australia in 1870, and New Zealand four years later. Today the wanderer is common at many places in the East Indies, on the Malay Peninsula, and in Burma. Its caterpillars thrive wherever their food plant is well established.

The arrival of insect immigrants, with or without human help, took on new perspective during the late nineteenth century as technology provided better transportation for people, and animals too. Scientists began to notice how the local numbers of any kind of creature rise with each increase in birth rate or immigration and fall with mortality and emigration. A population of insects might remain steady and inconsequential until some year when, for no obvious reason, the natural causes of death diminished. Afflictions from weather, diseases, parasites, and predators no longer held down the population. Only emigration could stave off starvation. The surplus individuals would only perish by staying where their parents lived. Perhaps emigrants could become successful immigrants elsewhere, starting a new colony where space and food allowed.

Examples of one-way travels, repeated at irregular intervals, turn up even on large continents. In the southern United States, an epidemic of caterpillars occasionally develops, followed before autumn by a snowstorm of adult moths. The particular species may be kin to a cutworm, with no common name at maturity. One with this special history is a dull brown adult with a single white point shining on each front wing. A fertile female moth contributes seven hundred or so eggs in any year. She lays them in rows low on grass blades, beginning in February near the Gulf of Mexico, as early as June in more northern states. A multitude of caterpillars emerges, each one grayish with three lengthwise yellow stripes above and a wide greenish-yellow band along both sides. A robin or a flicker may relish a few young crawlers, pecking them from among the grass roots where they hide all day. We like to believe that skunks and owls account for more at night, when the caterpillars feed exposed on low vegetation. Yet many grow to a length of two inches.

The name "army worm" seems earned by these caterpillars in the infrequent years when the amount of foliage the larvae can find at night fails to satisfy their appetites. Then the insects begin to spread out by day. Advance parties seem to be everywhere at once. Thousands of individuals march across roads, generally some in each direction, but conspicuously more in one. Countless numbers get crushed. The survivors travel on. Enough pass the critical point in growth that triggers a change in behavior. Without pausing to spin cocoons, they pupate naked on the soil. Soon moths emerge. The emigrants wing northward, some as far as Hudson Bay. Others react to cool weather by hibernating, either as full-grown larvae or as adults awaiting their turn to travel.

Continuing winds commonly provide the power that makes mass travel so spectacular. The timing of the flight then determines the destinations more than the size of the insect, so long as it is vigorous enough to stay airborne. "Aerial plankton" is the name given to such winged emigrants by C. G. Johnson, who began studying wind dispersal of insects in 1945 at the

Rothamstead Experimental Station in Harpenden, England.
They are wanderers, able to direct their travels only by choosing occasions to fly when the wind is likely to maintain a suitable heading.

A tiny aphid or a moth with a half-inch wingspan can get
swept onward by merely fluttering continuously in a breezer,
perhaps day and night. Small diamondback moths from northern Europe achieved a spectacular emigration in this way late
in June 1958. Meteorological data explained it later. One great
air mass passed Leningrad on June 28, combined with another
that had swept through a day earlier from slightly farther
south, and carried billions of the insects to the east coast of
Scotland on June 30 and July 1. A third air mass, overlapping
the previous two along their northern edge, came a day later
and picked up further travelers, funneling them westward. A
spectacular shower of the little moths settled aboard a ship on
July 4, a thousand miles west of Norway. At this pace the air
masses would have brought hordes of the emigrants to the
shores of the New World by July 6. But if any of these insects
completed their journey at this time, no Canadian scientist
recorded their arrival. In North America the species would
earn few headlines, for it has been there as long as anyone can
remember. Probably previous immigrants settled in. The
moth finds members of the mustard family for its caterpillars
to eat no matter where it travels. Mostly it depends on weeds,
unless people have planted rows of healthy cabbage or cauli-
flower.

Even large grasshoppers, such as the one famous since bibli-
cal times as "the" migratory locust, let low-level winds carry
their immense swarms from place to place. This behavior quite
often serves their ecological needs because the surface air in
which the insects fly converges with a cold front higher up and
coming from the opposite direction. The collision produces
rain, even over the deserts of Africa and the Near East. Rain
stimulates the locusts to breed and aids in the hatching of their
eggs. It causes spectacular growth of vegetation, on which the
hatchlings feed and develop.

Mistakes are frequent. The best educated guess of a weather
forecaster carries with it no guarantee that some overlooked
factor will not change the atmospheric pattern. Nor does the

leader of a billion grasshoppers preclude disaster once the message to leave spreads. Only a particularly vigorous locust may sustain its wing action for sixty consecutive hours. This may suffice for the insect to be carried forty-five hundred miles from the Sahel, south of Africa's great deserts, to England. Most of its fellows perish at sea along the way.

Several different grasshoppers of the Old World are celebrated as migratory locusts. Actually they follow no regular schedule, as true migrants do. Rather they change from almost solitary insects into gregarious ones and engage in mass emigrations from time to time. Their common features then are "locomotory drive," which starts them off on prolonged and undistracted flight, and certain structural features. These include longer-than-normal wings, effective muscles, and, above all, an inner reserve of food to provide energy for continuous fluttering. The urge to associate with others of their kind and to travel as a group seems the most challenging to explain, since it arises in some one year among insects whose immediate ancestors for several generations have been unsocial.

The most widely dispersed of all grasshoppers is the famous insect known to scientists as *Locusta migratoria*. Five different varieties inhabit grasslands throughout Africa, much of Eurasia south of the spruce-fir forests, parts of the East Indies, and tropical Australia; they range as far as New Zealand. Those that arrived like a devastating tornado on the land of the pharaohs, as the "Ninth Plague" predicted by Moses when he was eight ' years old, must have been no less fearsome than earlier and subsequent swarms. They descended on Egypt and devoured every bit of green vegetation. And when the "strong west wind . . . took away the locusts," as described in Exodus 10:19, the principal observation missing from the account was the wind that brought the plague in the first place.

The phenomenon defied understanding until 1921, when B. P. Uvarov in London explained certain events he had witnessed as a Russian in the Caucasus. The plague grasshopper in any country is sluggish as long as it is essentially alone. In a normally dry year only a few hatch from each batch of eggs, by absorbing enough water—an amount equal to their own weight—from dew that penetrates the soil where the eggs are concealed. A 'hopper rarely hops. It creeps, instead, up desert plants, and there feeds slowly. Its color is highly variable, and often matches its background so well that animals hungry for insects do not detect the locust. Its head grows large, its legs long, the thorax in front of its relatively short wings rather narrow with a high crest. As an adult, the insect in its solitary phase may find a mate and deposit a batch of eggs (or father them), or it may fail even in this.

Then the variable pattern of weather brings locally a few years with more rain than average. More eggs hatch. More maturing 'hoppers find mates. The population increases because of the higher birth rate. Little locusts see others of their own kind and are stimulated to greater activity. Their body pigments darken to a more uniform black-and-orange color pattern. The pigments absorb more of the sun's heat and the body temperature rises several degrees above that normal in solitary grasshoppers. The thorax grows broader, more saddle-shaped, with bigger internal muscles. The wings attain greater length and spread. Yet even before the insects mature, the food supply may give out. Too many active grasshoppers are trying to satisfy their hunger, to sustain their growth and their output of energy as they move about. Inadequately appeased, the insects begin to march. A few start out, and more join in. This is the last chance to stop them on the ground. One more molt will give them wings, and the swarm will take off into whatever wind is blowing across the open land.

Once a locust swarm is airborne, the only thing capable of stopping it is a change in weather. The pests fly on if their food reserves permit, as long as the temperature by day remains between 77° and 104° F., or above 81° at night. Greater heat or chill exceeds their inbuilt tolerances, and they alight to seek shelter; a downpour of rain has the same effect. Minor gusts of wind within the general flow of air have little consequence during a mass flight. Each insect is well fitted to compensate for any forced turns. The insect nervous system connects suitable sense organs to the muscles that drive the wings and makes corrections without the instrumentation an airplane pilot regards as essential. The actual heading in calm air seems of minor significance for, within a major swarm, actual groups shift in various directions for minutes at a time before adjusting (or overcompensating) to remain with the majority.

Uvarov's discoveries led the British government to establish an Anti-Locust Research Center in London in 1921. A network of field scouts came into existence in forty-two countries, to report on concentrations of locusts and changes in weather that might cause swarming of these insects. Control teams were dispatched with truckloads of poisoned bran wherever the locusts turned gregarious. Cheaper insecticides, available first in the 1940s, were stockpiled at strategic points, ready for distribution on short notice. The Food and Agricultural Organization of the United Nations began in 1960 to offer additional coordination, including tracking service for locust swarms both from light aircraft and from weather satellites.

Just once have we glimpsed huge swarms of these locusts in full flight. The pilot of an airplane drew our attention to them as he postponed his descent toward a landing at Addis Ababa, Ethiopia. Already a few high-flying locusts had splashed against the windshield of the cockpit at twenty thousand feet altitude. The rest flew much lower, in great rust-brown clouds. Hundreds of tons of insects comprised each swarm. Reputedly one ton of locusts consumes daily as much as 10 elephants, 25 camels, or 250 people do. Wherever they settle to feed, they strip the vegetation clean. The consequences for many centuries have been famine in enormous areas of central Africa and southern Asia, from Morocco to India, and from Tanzania to Turkey. No wonder people in these countries have so long feared the unheralded descent of a swarm as a plague, an "act of God."

The year 1969 was the first that the weekly "situation summary" from London listed no menacing locust swarms all summer. An atypical lack of rainfall over the whole region received the credit for keeping the locusts solitary. No accomplishment of humankind has succeeded so well in modifying their behavior. Yet much has been learned and used in designing control measures, regarding the factors that convert the insects from solitary to gregarious activity and back again to slugglish solitude.

Essentially the same pattern appears in all five races of *Locusta migratoria,* in its counterpart *(L. pardalina)* in southern Africa, and in accounts of the "lesser migratory grasshopper" or Rocky Mountain locust *(Melanoplus spretus),* which wreaked

havoc in the prairie farmland of the United States and Canada
during the 1920s. Indeed, the South African pest was the first
to demonstrate the key role of agitation in the development of
migratory body build and behavior. In the laboratory of the
Division of Entomology operated by the federal government
in Pretoria, J. C. Faure raised small numbers of these insects
in solitary confinement. He motorized the cages containing
some individuals to provide constant rocking movements. The
other cages remained quiet. Locusts that ate and grew in stationary cages matured as the solitary form, whereas those that
were forced continuously to adjust their balance in the rocking
cages all developed the characteristics of the gregarious
hordes. Disturbance, whether by adjacent locusts in freedom
or by machine, alters the internal chemistry of the 'hoppers in
ways that have promoted the survival and spread of the species
in past millennia.

Emigrating grasshoppers may meet especially capricious
weather in the Rocky Mountains, which can create "grasshopper glaciers" such as the one near the small town of Cooke,
Montana. Gordon Alexander of the University of Colorado
directed us to this special site where, each summer, the warm
sun melts the steep low end of the glacial ice. Thousands of
locusts thaw out—millions in some years. They tumble down
and decompose, exposed to air for the first time in many
centuries. Long ago those insects rode the winds upslope and
settled on the snowfields that feed the glacier. Too chilled for
further locomotion, the insects accumulated to a depth of
several inches, only to be covered and crushed by subsequent
falls of snow. The locusts became dark strata in the ice. The
glacier carries them all down the mountain slope a few feet
each year. Now scientists can gather newly exposed insects and
learn, by the radiocarbon method, that some of the bodies
perished under snowfalls fully six thousand years ago.

In a windy world, where small areas of vegetation are largely
surrounded by inhospitable peaks, deserts, or seas, an insect
of modest size may gain more from abbreviated wings than
through extended flight. Wings too short to use in airborne
travel can keep an insect walking or hopping with greater
security. Arid areas of our American West offer a perfect example: a plump cricket that sometimes appears in countless

numbers far from any coast. An outbreak of these insects threatened to destroy the first crops of grain planted in Utah by the Mormon pioneers. Every able-bodied man, woman, and child among the sixteen hundred colonists strove valiantly to save their food supply from an apparent repetition of the Ninth Plague visited upon the ancient Egyptians. Then came a "miracle," as though in answer to the multitude of desperate prayers. California gulls by the hundreds arrived, gobbled up the crickets, and saved the crops. The grateful Mormons erected a statue to honor the gulls and commemorate the sudden change in the fortunes of the settlement.

The gulls and the "Mormon" crickets still breed in Utah and many other states. We encounter a few of these insects on almost every trip through the area, and marvel at the shy

Emigrating Mormon crickets try to get past a metal barrier erected to stop them where they can be killed to protect crops. (Photo courtesy H. F. Thornley)

behavior of the crickets when they are not engaged in mass emigration. At a distance of twenty feet, the two-inch creature sees a person coming. It dodges first to the far side of the plant stem to which it clings. If we approach within ten feet, the insect backs down to ground level or lets itself drop, then creeps under denser cover. Only by sitting down ourselves and remaining motionless for a quarter of an hour or more are we likely to witness a resumption of normal activity, perhaps a meeting between two "Mormon" crickets in an open place.

Seldom today can we find a marching throng of these insects in order to observe how unwary they become when many others of their kind are in sight, even shoulder to shoulder. Now human observers keep watch over the sparsely settled western lands. They detect and eradicate any burgeoning population of "Mormon" crickets, migratory locusts, or other conspicuous creatures that might attack the crops or range plants. Rarely does an infestation develop to the stage where barricades are needed to stop its onward march or to slow it until insects climb over other insects and finally pass the barrier.

Concerted efforts by humankind each summer minimize the impact of would-be travelers. The Rocky Mountain locust may now be extinct, never again to add a layer of chilled bodies toward perpetuation of a grasshopper glacier. Other potential emigrants are merely held at bay.

Island insects are equally vulnerable. Even to flutter upward from the ground is to invite being blown out to sea and drowned. Many a grasshopper, beetle, or bug with a long line of ancestors on the same island reveals one inheritable reaction to this frequent hazard: they are flightless through hereditary failure to develop usable wings. On the islands of the Galápagos archipelago, straddling the equator some six hundred miles west of Ecuador's Pacific coast, we find four out of the six kinds of grasshoppers to be flightless, all members of the indigenous genus *Halmenus.* Many beetles of several families possess only reduced wings there. The grasshoppers, at least, can jump. But no grasshopper leaps as often as most people expect. It uses its oversized hind legs to jump from any danger, but creeps or walks on its four forward legs from the day it hatches for most of its life.

Many a conspicuous insect, particularly in the tropics, keeps secret the long-term values its species finds in conspicuous behavior. We have appreciated this on many occasions since the day when two swift-flying insects rushed past us as we descended the gangplank, debarking at Cristobal near the Atlantic end of the Panama Canal. They were followed by a third, then half a dozen more. While we waited for our baggage, a seemingly endless stream of the same species hurried by, as though rushing to some important destination. Gradually we glimpsed enough to gain a clear picture of the flapping creatures: three- to four-inch wingspan, conspicuous dark tails from the trailing edge of the rear wings—butterflylike wings brilliant with metallic green intermingled with narrow areas of jet black. The insects simply had to be day-active moths, members of the tropical genus *Urania*. We had read about them. Now we were in the midst of one of their inexplicable emigrations. This one, at least, led only to watery graves, for no land lay ahead of the moths closer than the opposite shore of the broad ocean.

We have witnessed similar flights from the deck of a ship passing through the Panama Canal, particularly in June or July. The dazzling sight makes us recall Charles Darwin's experience aboard H.M.S. *Beagle*, fully ten miles off the bleak Patagonian coast of South America, as he recorded it in his journal for December 6, 1833—early summer in that part of the world:

> One evening . . . vast numbers of butterflies, in bands or flocks of countless myriads, extended as far as the eye could range. Even by the aid of a telescope it was not possible to see a space free from butterflies. The seamen cried out "it was snowing butterflies," and such in fact was the appearance. More species than one were present, but the main part belonged to a kind very similar to, but not identical with, the common English [Clouded Yellow]. . . . The day had been fine and calm, and the one previous to it equally so, with light and variable airs. Hence we cannot suppose that the insects were blown off the land, but we must conclude that they voluntarily took flight. . . . Before sunset a strong breeze sprung up from the north, and this must have caused tens of thousands of the butterflies and other insects to have perished.

Curiosity may be aroused for anyone on shipboard or at a dock, but there are few first-hand answers to the many questions that come to mind. What gives the fliers their uniform heading? Does it differ from one day to the next? What proportion of the total insect population behaves in this self-destructive way, and how many remain at home where they renew the population?

Even on land for an extended stay amid a mass flight of butterflies or moths, we can count and examine the travelers more easily than we can understand their actions. Allen M. Young tells us how he took a census of *Urania* moths in various places in Costa Rica. He accomplished little else for forty-two days between August 12 and September 24. He began at sunup and stopped at sundown in each of three localities, at elevations from 125 to 4,500 feet above sea level. For five consecutive days at each site, he counted the spectacular traffic, then moved to the next place in a regular rotation. He tallied the fliers that passed between two trees about fifty feet apart—where he could see individual insects against the sky—regardless of whether their travels were five feet or eighty feet above the ground. This small sample gave him a total of 54,625 individuals. Their heading remained closely the same regardless of gentle breezes. They flew at speeds between $1\frac{3}{4}$ and 2 miles per hour, mostly without pausing anywhere.

The furiously flapping swarm of *Urania* moths includes members of both sexes. Apparently they have suspended simultaneously their hunger and sexual drives until the survivors can reach some destination. Toward nightfall, the fliers increase their speed and travel as far as possible prior to settling on treetops for the dark hours. Before rushing away the next morning, many of the moths visit the small white clustered flowers of a mimosalike tree (a local species of *Inga*) in the forest canopy. Probably the insects take this opportunity to renew their water supply from nectar and dew. By 7:00 A.M., almost all are airborne and on their way. So far as Young could learn, the females are all unmated, and four-fifths of them bulge with developing eggs.

To anyone whose experience has been mostly outside the tropics, it seems almost incredible that so common a moth should be unknown in the caterpillar stage, and its food plant be a complete mystery. In the rain forest, this situation is understandable because the dedicated scientist tends to be on the ground while most of the vegetation is high overhead in the dense canopy of interlacing limbs, twining vines, and perching plants. Somewhere in this tangle, *Urania* caterpillars must live in spectacular abundance, offering no recognizable sign below of their presence for about ten months each year. Perhaps the maturation and emigration of the moths corresponds to a hidden effect of the dry season far above the ground. Someone will surely learn the answer and simultaneously uncover more intriguing mysteries, for that is the open-ended way science progresses. Meanwhile, the insects continue extending their heritage wherever they can, for as long as any versatile life has a place on our planet.

7

Insects
in Social Groups

BABOONS, PENGUINS, AND WEAVER BIRDS SHARE A COMMON FEA-
ture with ants, termites, and some of the bees and wasps. All
need close association daily with others of their kind. Not one
of them can survive for a normal lifespan as a hermit. Their
future is predicated upon social behavior.

For hibernation, ladybird beetles assemble in spectacular
numbers. They gain from inactive togetherness in a simple
way. Each insect releases a minute amount of safeguarding
repellent, and the vapor from so many warns off any predator
that might not detect the scent from a solitary beetle. The
same safety system is used by the heliconiid butterflies that
assemble for social roosting overnight, then disperse after the
sun has caused flowers to open.

An alert parent is more active while tending eggs or young.
Yet the old explanation (that the parental behavior—motherli-
ness and fatherliness—is called forth entirely by hereditary

internal guidance) seems simplistic today. We look for a more tangible reward in the dark privacy of an earwig's nest, where at least one parent stays curled around the eggs until they hatch. For a while the hatchlings stay close to an adult, although old and young must eat unlike food because of differences in size. Close study often reveals that the adult takes some attractive exudate from the eggs or young; the juveniles stay with the parent as a traveling food source from which they solicit repeatedly.

Social treehoppers may be different again, for both young and adults stand separated from one another, in plain sight, each with sucking mouthparts thrust like soda straws into the same plant stem. These insects may not touch even when they move about to try a new location. Yet Thomas K. Wood of Wilmington College in Ohio finds statistical benefits to the immature treehoppers from associating closely with adults. Predators, such as birds, seem less likely to attack the group

A mother treehopper and her young cling to a small branch. As long as the parent is in attendance, fewer insect-eaters attack. The gain for the young is clear, but it is not known what the mother gains by staying close to her young and competing with them for plant juices.

as long as the somewhat larger parent is among the smaller young. "Safety in numbers" is not the answer, since the birds attack the juveniles if the parent is removed. What does the adult get from staying with her brood? The reward must be real, since the mother competes with the young (and they with her) for plant juices. Surely such continuing sociability is maintained by some tit for tat, a quid pro quo, one service in return for another.

Surrogates serve many insect societies—as well as humankind. Usually we dignify them with words that suggest their separate roles: workers, soldiers, and the like. A queen, primarily a reproducer, may labor at first as a construction worker, a food gatherer, a brood tender. Later the workers will free her from these and countless other tasks. Workers may serve in place of specialized soldiers, which commonly provide a civil defense force rather than the ultimate warriors. It is workers that become raiders and slave-makers on occasion. The nearest approach to a sedentary job is held by certain worker ants known as "repletes." They accept liquid nourishment from food gatherers, and store it in a distended crop— an extraordinary organ that swells to spherical shape. Until relieved of this reserve by other workers, the replete clings to the wall or roof of a subterranean chamber as a living reservoir.

The success of insect societies depends upon the interplay of individuals as they follow diverse patterns of behavior. These vary according to age, to sex, to temporary or long-term role, and to the heritage carried by all members of each species.

CUES FROM THE YOUNG

The marching ants in Ecuador follow a nomadic routine, setting out at dusk and settling soon after midnight in a massive bivouac. Twenty to thirty million individuals may cling together, suspended below some rain shelter. At dawn they shake themselves out of this amazing cluster and go hunting. (No sightseeing for these ants, because each of them is almost blind.) For five weeks at a time, the workers issue from the camping site to overwhelm the small animals that serve as food. Then they are ready to move on again. Workers file out, many of them carrying grublike larvae, followed by the heavy-bodied queen, who is assisted by a crowd of workers.

Only once have we been so fortunate as to witness the arrival of a column of army ants. We had returned about an hour before lunchtime to our combined laboratory and bedroom in a small cottage on Barro Colorado Island in Panama. We were busily working at a table with some specimens brought back from a morning field trip. Since the day was fine and we had learned not to fear any of the insects that might flit from the sunny hillside through our open door, we gave no thought to being invaded. A slight rattling sound made us glance toward the narrow porch. There we saw a wave of ants swarm up over the edge and along the floor. Their approach gave the same feeling of unreality as a movie run backwards.

We covered our specimens with glassware and hurried to kneel as close as we dared to one side of the advancing ants. Incredibly, they had no leader. At the oncoming edge of the swarm every ant faced its fellows, but was being pushed backward reluctantly into unmarked territory, unable to resist the momentum of the oncoming throng.

Along the sides of the procession we saw larger ants, fully three-quarters of an inch from abdominal tip to the ends of their hooked, down-hung jaws. Pale-headed, with a minute eye each side, they clambered along as close as they could to the smaller, small-jawed workers without getting their legs tangled. These soldiers ran short distances forward, then back a few steps, and onward again. The English naturalist, Thomas Belt, described them as officers, "directing the columns," but in fact, they serve as defenders.

The onrushing ants reached a sealed carton resting heavily on the floor. With their backs supported by the carton, the ants along the leading edge could now resist the throng that shoved against them, and the column divided. A third of the ants went around one side of the carton, two-thirds around the other side. They met and fused into a single column, bypassing the box as they might the trunk of a tree. Soon some of the ants swarmed up over the carton, explored its sides and top, and descended to join their fellows. We set a banana, still in its skin, atop the box while the ants were there. They covered the banana in a few seconds, made no attempt to bite into it, and then ignored it. Apparently they left their scent on its surface, just as they did on the carton, and oncoming ants wasted no time on further exploration.

The army flowed on its myriad legs, rattling fallen bits of paper, investigating every corner and crevice, marking everything with a faint odor that initially kept the column in formation and then prevented it from going the same way twice. By lunchtime, the ants had been everywhere indoors on floor, walls, furniture, window frames. Not a trace remained of a dead fly on the window ledge. It had been pulled to pieces and the edible fragments sent back along the column and the column's trail to the nest. The ants cleaned everything and molested nothing. We had closed up all our valuables. The insects did not take time to pry.

After lunch we found the column again and traced along it to its advancing end. In two hours it had traveled at least fifty feet. Now the front was broader, and the hunting more productive. On the ground the ants were swarming over beetles, tropical cockroaches, tarantulas in their dugouts, even grasshoppers and other insects that we fully expected to fly out of reach. Flight, actually, may rarely have survival value. Too many "ant birds" follow the army ants all day, and pounce upon any potential prey that takes to its wings.

Most ants of other kinds are easy victims. But a few kinds have a system for saving their own lives and those of young— either grubs or pupae. These ants rely on the normal behavior of army ants, which rarely climb a blade of grass or a slender twig. To be fairly safe, these other ants hold their young gently in their jaws and back up a grass blade, leaving no trail to invite an army ant. Soon the swarm will have passed. The peril will be over. The insects that hid in plain sight while the army marched through can return to their homes.

The queen of the army ants remains with her brood in the bivouac all day. Returning workers bring food to them, following trails from the advancing front, each ant with a small piece to carry. Not until dark will all ants set out for a new campsite. Yet these moves by night and hunts by day reflect a basic restlessness of the workers and soldiers. It is renewed each time one of these sterile individuals gets close to a larva that is half-grown or more, or to an adult that has recently emerged from its cocoon and is still pale in color. These members of the colony secrete a special exudate, which is licked eagerly from their bodies. Originally, this attention was included in the behavioral category of parental care, even though the workers are substitute parents at best. Now the exudate is recognized as an important communication between young and old. It programs the whole colony in an alternation between nomadic and stationary phases.

As soon as the larvae spin cocoons in which to transform to adults, the exudate becomes unavailable. The workers and soldiers go on strike. They stand around. They avoid daylight and seem to sleep all night. They leave the queen and the cocoons in the bivouac, which is ideal for the queen. Recently she has been experiencing difficulty during each nightly move because her body is so swollen and heavy with developing eggs. Now that she can remain quiet, she begins laying. Her output may rise to almost an egg per second—over eighty thousand in a day! Gradually her body shrinks. She loses weight and becomes more agile. Meanwhile her eggs are hatching, and workers in her vicinity receive enough stimulating substance by licking the tiny larvae to spend time caring for them. At the same time new workers begin hatching from the cocoons. The older, darker workers get excited as they clean the pale, callow individuals. Before long the colony seethes with activity. The stationary phase ends. The army of workers goes hunting by day, no longer repelled by light. Capture of food and the interchanges that occur while sharing it add to the animation. The whole population will move to a new bivouac as soon as darkness comes.

The chemical cues from larger larvae and from young adults coordinate many different aspects of behavior in the nomadic ants. Identifying the separate effects, such as the change between shade-seeking and light-tolerating, has become a research specialty for Howard R. Topoff, a psychology professor at Hunter College of the City University of New York. Topoff seeks to analyze the seeming determination of the marching army ants. In these endeavors he is extending the work of Theodore C. Schneirla of the American Museum of Natural History. Like Schneirla, Topoff camps close to army ants wherever he can find them. He crawls on hands and knees through undergrowth, often with difficulty because of sharp thorns and spines, to examine every aspect of activity in the surging columns.

Observations at close range are less risky than might be anticipated. The moving mass of ants can be kept in view without fear that individuals will discover the onlooker and send out an alarm. No army ant goes hunting by itself, although it is strictly a carnivore. The group acts as one, almost in concert. Every female, however, whether worker, soldier, or queen, goes armed with a potent stinger. This weapon is celebrated in the name *Dorylinae* (for the subfamily of ants to which the army [and driver] ants belong), from the Greek word for spear. The soldiers have their hooked jaws as well, ready to bite—but not to release, even if they are decapitated.

We would take no chance on having a wave of these insects sting and bite us. Yet the broad effect of these creatures of the forest floor may not be lasting. Schneirla found only about fifty colonies of army ants in any one year on the sixteen hectares (forty acres) of Barro Colorado Island, isolated by the waters of Gatun Lake and close to the interocean route of vessels using the Panama Canal. On hunting days, the food-gatherers may travel no more than two hundred yards from the bivouac of the night before. Their impact is reduced to 50 percent by their slow alternation of hunting and resting (egg-laying and pupal) phases. Only where the workers and soldiers march is their influence so commanding.

The tropics of the New World have no monopoly on nomadic ants that travel in predatory streams. The army ants of America—about 150 species—have a few outliers in the southern and mid-central United States. Their counterparts—another 100 species—live in warm parts of Africa, Asia, Indo-Malaysia, and Australia. Those of Africa are called driver ants. They were the first insects we encountered on the Dark Continent—on the nearest tee of the golf course across the road from our hotel in Entebbe, Uganda. "How appropriate!" a golfing friend exclaimed when we mentioned this encounter. But driver ants get their name from no club in the golfer's bag. They sting and bite the feet of antelopes and other conspicuous animals on the African plains and start them running, just as though an army of beaters was driving the game past colonial marksmen—a familiar sight in an earlier era.

Our experience at that time gave us no way to know that these were driver ants *(Dorylus)* and not army ants *(Eciton),* except that they were part of the wildlife within sight of Lake Victoria. They closely resemble army ants in appearance, behavior, life history, and impact. But the drivers build larger colonies, with more members in the family than any other kind of social insect. Occasionally the workers and soldiers number in excess of 20 million, all coordinated chemically by subtle substances from their young. The whole lively colony may weigh twenty kilograms (forty-five pounds).

Among such hordes, it seems impossible that one individual could ever recognize another as a unique ant, one that has been met before. We can manage this by daubing harmless spots of colored paint upon the backs of twenty ants, or fifty, or a hundred. Beyond this our marking system fails, and all unmarked individuals remain anonymous. Yet our crude technique allows us to satisfy our curiosity on one point: no worker ant concentrates her attention on a few larvae to the exclusion of all others. "This baby is mine to care for" implies a division of responsibility without which social insects manage effectively. The whole appearance of mother love, whether by the queen with her first small brood or by her sterile daughters with later hatchlings, is an uncritical human interpretation with no more depth than the recognition of a person's face in an inkblot during a Rorschach test. The care provided is a pattern of response to flavorful secretions that the young individual produces and the adult seeks. Each chemical is a cue, a one-way communication, but also a system of quid pro quo. The larva grows with the meal, while the worker goes off excited by the exudate.

A forest of tall trees, holding foliage high to the energizing sun, can make us feel like ants ourselves. In a tropical rain forest, the lowest branch may be ten meters (more than thirty feet) above the ground we walk on. Anything as small as an ant is invisible on the leaves above our heads. Yet ants live out their lives in the forest canopy, never descending unless the tree topples. Some chew up plant fibers to make strong, paper-thin nests on overhanging limbs. An anteater can rip these

structures apart, but birds and most reptiles are fenced out.
Flimsier, yet far more extraordinary, are the leaf nests constructed by weaver ants in the treetops of Africa and tropical
Asia.

The favorite nourishment for weaver ants comes in a multitude of living packages and in almost any tree. Insects of other
kinds are killed and eaten piecemeal by the larger of the two
sizes of workers, which patrol the stems and leaves. The same
workers react quite differently when they discover aphids
(plant lice) or scale insects. These sucking insects exude droplets of excess sugar to balance their diet of tree sap. The ants
drink the droplets, share the sweets with passing workers, and
stand guard over the supply. All the running about is by the
larger workers. They regurgitate food to one another, to the
smaller workers (which tend the eggs and younger larvae), to
the larger larvae, and to the queen. At frequent intervals a
large worker near the queen extrudes a minute infertile egg,
which the queen accepts as a special snack. Her majesty and
her young remain totally incapable of feeding themselves.

The larger workers and older larvae have a different task to
perform together almost every day. The procedure begins
soon after dawn as the workers explore new foliage or
branches of the tree that the colony has not yet claimed. Individual workers begin struggling with the edges or the tips of
leaves, curling the plant tissue upward wherever it will yield.
The concave surface, as contrasted with flat portions of the
leaf, attracts other workers. Soon the ants are lined up and
pulling in unison. Or they hold to one another in chains while
still more ants clamber on their backs and struggle to shorten
the chains. Leaves that were parallel become sides of irregular
tents, all held in shape by the tugging insects. How long can
they cling? Many workers rush about, clearly intent on what is
being accomplished, yet finding no place to get a useful jawhold or clawhold.

These larger workers are no circus roustabouts, well practiced in raising a carefully cut canopy of canvas. They are
half-inch insects, struggling with whatever leaves and
branches the tree provides. They get no guidance or encouragement from an experienced, keen-eyed director. Half the
labor force, in fact, runs around to no gain. Yet the workers

that have found no place to pull suddenly go racing back to that part of the nest where the larger larvae are. Each of these workers soon returns with a squirming larva securely gripped in tonglike jaws. To the gaps between one leaf and the next these workers run. Now they swing their bodies vigorously, wave and squeeze the larvae. Each move is exquisitely controlled.

The mouth end of the larva touches briefly on one leaf at the beginning of a swing, then the opposite leaf at the end of the move. Every touch applies a small drop of secretion from minute gland openings just below the larva's mouth. The exudate is liquid silk. The swinging action stretches a fine strand of it to the opposite point of attachment. Soon thousands of strands bridge the gaps between the leaves. They adhere and dry with amazing speed and strength. A weft of fibers becomes a silken sheet, woven by the ants and their larvae to an exact fit. The workers fashion circular entranceways and silken walls of inner galleries, then carry the larvae deep in the nest again, to be cleaned and fed until they are fully grown.

Tactile and olfactory signals must guide the ants as they perform these extraordinary feats of engineering. Few other avenues of communication are open for the eyeless larvae, which serve as tools when the irregular nest structure approaches acceptable form. Yet all weaver ants mature with larger eyes than is usual in members of other subfamilies. Vision must aid them, too. They not only remember the form of their nest and all its passageways, but notice if some object near the nest is moved without touching it. The object might be another branch from an adjacent tree that a scientist bends until it almost touches. The ants will cluster as close as they can to the object beyond their reach. They climb atop one another to produce a rough pyramid of bodies and legs. If the topmost worker can reach across and make a living bridge, more workers race over to the new territory. In an hour the ants may have brought larvae to weave a silken span into a firm highway for workers going in both directions.

The weaver ants seem more like ourselves than most ants do because they spend so much energy in controlling their environment. They expand the silken shelter that an ant larva spins for a private cocoon until the woven strands provide security for the whole colony in an entire treetop. Within the tremendous nest, the workers weed out (and eat) any insect or other animal that is not useful to them. Curiously, some weaver ant species tolerate the caterpillars of certain small moths, as though judging correctly that the caterpillars will contribute more than enough extra strong silk to the ant nest to compensate for eating a few small larvae from the ant brood.

People from the western world are just now beginning to find uses for weaver ants in horticulture. The Chinese have been collecting, marketing, and employing a widespread Asiatic species since at least A.D. 300. Whole nests are cut from trees of no commercial interest, and tied to the branches of citrus trees. The ants accept the new home and keep it free of most insect pests. This most ancient of biological control methods proves useful in Ghana in orchards of cacao trees. The ants even control the spread of virus diseases by attacking the sap-sucking leaf bugs that transfer the infection.

We can think of a weaver ant as employing child labor when it swings a larva back and forth like the stitching needle of a sewing machine. We envy the ant because the "needle" needs no threading; neither the spool nor the bobbin run out of material in the middle of a job. The larva must merely be returned to the brood section of the nest, to be recharged like a nicad (nickel-cadmium) battery for the next operation. The larva, in fact, invites these attentions as part of its role in the colony. It never breaks or wears out. Instead, it progresses to become a worker—and if one of the larger size, to use subsequent larvae from the same parental line in the corresponding way. What wonders the genes call forth in structure and behavior!

While watching a weaver worker wave a larva back and forth, it is easy to forget that the larva is no inanimate tube of cement. It is a cooperating individual, one that will mature to take on many different tasks. This role in producing silken threads is merely a digression from the usual business of taking food, signaling for more, and growing. Within the nest, the workers respond generously to the hungry larvae. Yet the signals from the young as they solicit this service are so inconspicuous that we can scarcely detect them. A minute droplet of odorous exudate may suffice, bringing workers with nourishment along dark galleries to the place where the hungry larva has been left. Location alone cannot serve, for ants prepare no brood cells or chambers for the young comparable to those constructed by wasps and bees. Instead, as has been observed for centuries, the worker ants move their larvae from one chamber to another inside the nest at frequent intervals.

An ant nest in the ground is more familiar to most people than one in a tree or a temporary bivouac of nomadic ants that have no fixed abode. Subterranean ant nests, where the workers tend their queen and young, may bulge upward and become lasting features of the landscape. For more than a decade now we have refrained from disturbing a mound occupied by a big colony of the field ant on our property in New England. It is more than a meter (four feet) across and about a fourth as high above adjacent ground. In Europe, one colony of woodland ants has built bigger mounds that have neither enlarged much nor changed shape appreciably in better than fifteen years—the span for which photographic records have been kept.

The form of the ant nest gives little hint of the social interactions between young and adults in the maze of interconnecting passageways. Far more can be inferred by watching the workers as they enter the doorway with booty from their expeditions beyond the nest. Are they seed-eaters ("harvester ants"), or leaf-cutters? Do they return home full of honeydew, with nothing clamped between their jaws? Or are they predatory hunters, bringing insects and spiders in various stages of disrepair?

The leaf-cutters of the American tropics never fail to fascinate us. Their well-traveled trails catch our attention as we walk along, scanning for indicators of animal life anywhere between four and forty feet from our momentary location. An ant highway will be ten centimeters (four inches) wide or so, brown through a world of green, straight as many an unpaved street, and slightly depressed below the general surface of the ground. Years of traffic wear it down. By kneeling for a closer look, we learn which direction is toward the nest and which toward the leaf supply. Pale red ants a centimeter (three-eighths inch) long will be traveling both ways. Outbound individuals carry nothing. The home-bound workers keep their heads high, the better to hold pinched between their jaws a fragment of green leaf or a petal from a flower.

Some pieces of plant material seem too small to require a whole ant to carry them. Other burdens exceed their carriers in one or two dimensions. Perhaps a worker stumbles along, head twisted to one side, struggling with minibrute force to pull an inch-long blossom fragment. The load moves slowly, catching on everything, interfering with traffic in both directions. Or a fragment of thin leaf as long as the worker in height and breadth resembles a green sail, held vertically and almost straight above the back of the ant. So conspicuous an encumbrance distracts attention from the insect that transports it. Yet often the leaf fragment serves as a perch for a smaller worker of the colony—a "minim." This is no baby in a backpack but a working member of the team, one alert to fend off any small parasitic fly that might otherwise steal a free ride into the subterranean nest.

Often we discover the source of plant materials by following back along this parade of burdened ants. The visible highway ends, but the workers pay no attention because they have odorous trails to follow. With elbowed antennae tapping alternately like two canes in the hands of a sightless pedestrian, the insect detects smears of secretion left by previous workers along the route. Individuals returning nestward with a good load mark the substrate with a volatile secretion. Outbound ants follow the scent to the same or adjacent territory, and there explore on their own. If they find nothing, they retrace their steps but leave no substance that might mislead another ant. Each trail that becomes worthless soon evaporates. Successful explorers mark new routes back to the main highway.

If the leaf-cutters travel vertically on the smooth bark of a tall tree in the rain forest, we cannot follow them into the canopy to see them work. But if their trail ascends the person-high stalk of a big Caladium leaf, their operations become visible in the half-light of an opening in the jungle. In single file the workers scamper to the edge of the leaf blade. Each ant of modest size carries its own tools—its sharp-edged jaws. As though wielding a stencil-cutting knife, the ant slices through the leaf in a curve that frees a fragment, large or small. As the trophy begins to sag just before coming loose, the ant brings its jaws together. Adroitly it raises the prize overhead, out of

Leafcutter ants cut pieces from a big Caladium leaf (about three feet long and two feet wide) in Ecuador; in the photo opposite, this leaf has been cut until only the midrib and branch veins remain.

the way of those sensitive antennae. Now the insect must find its way back to the nest, even if reaching that destination takes the remainder of the day. A minim may climb aboard to guard it if the piece is large, but not until nearer home along the highway. By then the Caladium leaf will be merely a strong stalk with a fringe of branching veins along each side. The trail down the stalk will be fading fast.

Does a leaf-cutter ant speed up as it reaches the doorway of its nest? Or does the sand around the entrance hole allow faster travel, fewer encounters with ants going the other way? Before noon the outgoing traffic exceeds the inbound; later, the opposite is true. Probably few workers make more than a single round trip in a day. A dozen trips could be the labor of a lifetime. No ant of this kind comes out at night. A rain stops them anywhere, and often clogs the entranceway with wet landslides that must be cleared quickly to let every latecomer return to safety before sunset. Only deeper in the earth can the

insects labor at any hour, for there the larvae must be fed.

These ants are vegetarians from the day they hatch until they die. Yet none will eat any part of the plant material the workers haul home so laboriously. Instead, every fragment is minced and tucked carefully into a subterranean compost heap. The few that we have seen exposed when some excavator cut into a big ant nest reminded us in bulk of an old-style bushel basket (about thirty-five liters). Nothing in the mass seems dry or loose, for this is a fungus garden under critical control. The smallest workers (the minims) test it constantly. They and the larger workers wet it with their saliva, fertilize it with their feces, nibble at the fine fungal strands, and harvest little knob-shaped outgrowths that form the complete diet of every leaf-cutter at any age.

The young ants may resemble human infants in swallowing whatever food is put into their mouths. But the larvae grow to adults without passing through any rebellious stage. The tiny

knobs of fungus satisfy their hunger even after they mature
enough to tend the compost heap, or to journey great distances to cut tissues of flowering plants upon which more
fungus will grow.

The fungus-tending ants of America parallel closely the fungus-growing termites of the Old World tropics and subtropics.
They are agriculturalists, whose pattern of behavior from
hatching onward calls for the equivalent of a domesticated
plant. The ant colony sustains its inherited organization on
this one controlled resource. Yet, almost certainly, the ancestors of leaf-cutter ants millions of years ago were equally con-
sistent in relying upon a carnivorous diet.

All ants, entomologists agree, arose from insects much like
hunting wasps, and diversified spectacularly during late Creta-
ceous time, much as bees did in a totally different way. It may
be that ants went underground and into protected nests because predators and parasites pressed ever more strongly
upon any that failed to change in these directions. Today,
some of the most superficial nests are made by the most vigorous ants of all—the bulldog ants of Australia—which impress
Caryl P. Haskins as being the most primitive ants on earth.
Their larvae are also the most demanding.

Australians warned us about bulldog ants. These and certain poisonous snakes would be the most fearsome denizens
we might meet while field-tripping on their continent. No
snake came our way. But we did uncover several shallow nests
of the bulldogs when we displaced fallen logs and stones that
lay on soil. Each time the ants leaped in our direction, ready
to bite and sting. Since each worker was two centimeters
(three-fourths of an inch) long, with strong legs and stronger
jaws, we kept our distance. Yet our exposure of the nest cavity
let us see almost every time a little cluster of translucent eggs,
several piles of squirming larvae, and occasionally some big
cocoons. To our surprise, the belligerent workers made few
attempts to pick up their cocoons and young to haul them to
safety in deeper galleries. Instead, the ants sought to attack us
until we replaced their roof.

Bulldog ants hunt alone, and bring home both smaller and
larger insects as nourishment for their queen and her brood.
At the entranceway the worker may cut the prey into pieces,

the better to carry it below. Tiny pieces are dropped among the loose clusters of small larvae, bigger fragments or whole victims amid the looser groups of large larvae or before the queen. The larvae must feed themselves, for no liquid droplets are regurgitated for them. They squirm and probe with their mouthparts to break off and swallow particles of food. As supplementary rations, both the young and queen accept small sterile eggs which the worker females extrude at intervals, then proffer neatly pinched in powerful jaws. How dependent the queen becomes on her daughters to bring food remains uncertain. At least for a while after she founds a colony and excavates a nest, she alternates between laying eggs and making her way out to hunt to satisfy her hunger. No other ants are known to have such vigorous queens, or such enterprising young.

No other ants are the bulldogs' equal in quick movements, agile aggressiveness, or reliance upon vision as well as other senses. Were it not for the telltale enlargement of one abdominal segment between the hind part of the body and the thorax, we could easily mistake a bulldog ant for a wingless wasp—and a dangerous one at that. Actually, each ant has a more distinctive feature or a vestige of it: a special gland on each side of the third thoracic segment. Its secretion seems to render the ant immune from fungus infections, and probably explains why so many members of the ant family (Formicidae) thrive underground.

Aggressive wasps that socialize in colonies with as many members as cooperate in an ant nest are well known. Some of them appear almost as casual as a bulldog ant in bringing a succession of victims (either insects or spiders) for their legless larvae. These devour what they are given, as best a wormlike creature can. We should remember, however, that these young insects can open their mouths widely. Internally they are mostly stomach and short intestine. They are digestive machines. Surely they signal their mother, or a sterile sister (worker) instead, when she comes with food. What passes between larva and adult to keep the food collector coming?

Recent research points to a pattern that may be found much more widely when researchers examine in minute detail what goes on when full-grown insects associate closely with young members of their species. Jacob Ishay and R. Ikan at Tel-Aviv University raised this possibility while studying the wasp-waisted *Vespa orientalis* of Israel. This irritable hornet, which always seems so ready to sting repeatedly, hunts other insects and carries them to the nest. But neither the winged workers that do the hunting nor the queen that stays with her brood in the nest has the enzymes necessary to digest protein. The nutritional dilemma is solved through community action. The workers feed fragments of the prey to the larvae. The larvae perform the digestion for the colony. They use some of the food for growth, but regurgitate simplified nutrients, including sugars, to the workers and to the queen. In the privacy of the nest, protected by an armed force of hunters, her majesty lives exclusively on "milk" that she begs from her young.

Something similar is suspected in Death Valley, along the border between California and Nevada. There the harvester ants, which go forth to gather seeds when the weather permits, do no more in the nest than prepare the kernels to feed the young. The larvae alone have the capacity and the physiological ability to simplify such plant materials; they share the nourishment throughout the colony. Higher on adjacent mountains, the montane ants of forested land stay underground most of the time. No one is sure what the adults eat. But fungus strands, similar to those associated with tree roots, are accepted readily as food by the ant larvae. If the fungus is made artificially radioactive, the tagged atoms turn up inside the larvae within two hours, and subsequently inside the workers too. Probably the larvae do all the feeding, then regurgitate a predigested surplus to adults and keep the colony thriving. The cue from the young may be an invitation to dine.

THE LONG-LIVED QUEENS

The common feature most evident among the social insects, whether termites or ants or social wasps and bees, is a long-lived queen. She is incredibly prolific, and maintains the cooperation of other individuals whose work together perpetuates the species.

Naturalists for years have regarded motherhood as basic in each society. Unlike humankind, insects concentrate their so-

cial systems around a single mother, the impregnated queen.
She founds the colony and raises a new generation. Her first
offspring aid her in tending still more eggs and young, or take
over these parental duties altogether. Then the queen can
specialize in producing eggs. An increasing retinue of her
progeny then bring her food, keep her clean, and rush her
eggs to the nursery. Luxury for her, by dividing duties, promotes efficient growth and reproduction of the colony.

A special significance to motherhood among insects came to
light with the discovery that the queen exudes a "queen substance." It rewards and motivates members of her family who
groom and feed her. They share it among others in the colony
and thereby spread the basis for coordination. Without a sufficient supply of queen substance, the social behaviors change.
The extended family splits up, unless some substitute royalty
can come forward to assume the queen's role.

Closer study reveals that the queen is not so passive an
egg-producer as she appears at first glance. If she is not
equipped for controlling her associates chemically, with queen
substance as a tranquilizer and reward system combined, she
behaves more like a top executive. She attracts to her with
rewarding substances the most competent and vigorous of her
mature daughters, then bullies them physically into working
solely in her behalf. If any of these subordinates manages to
lay eggs as private enterprise, the queen eats them without
permitting a similar liberty to be taken with her own eggs.

These inherited programs appear to have served termites
and ants for 100 million years or more. Fossil insects of both
types appear in sedimentary rocks dating from mid-Creta-
ceous times. Some of the ants preserved so beautifully in Baltic
amber of Paleocene age—perhaps 65 million years before the
present—seem scarcely different from kinds alive today. No
one has reason to suspect that their colonial activities altered
more than their bodies during that immensity of time.

People who observe a termite nest commonly call the insects
"white ants," because of their pale color, great numbers, customary size, and scurrying antics. That a termite is waistless,
whereas an ant is constricted at the middle, is easily overlooked. Scientists tend to regard termites as supersocial cockroaches, and to credit these two types of insects with a com-

mon ancestry—one quite remote from that of the wasps, ants, and bees. This belief rests not only upon anatomical similarities but also the digestive habits of most termites and of certain cockroaches—the wood-eating roaches that do not scavenge like most of their near kin.

We have exposed these special roaches in the woodlands of Virginia, for they thrive there as well as in adjacent states and in similar habitats of eastern Asia. We caught the insects as they scurried into dark crevices when we pulled apart rotting stumps a short distance from the Mountain Lake Laboratory of the University of Virginia. At the time, we were not as curious about the primitive social life of these dark red roaches as we were about their digestive tracts. A compound microscope reveals that a wood-eating roach of any age has intestinal partners—single-celled protozoans with multiple long flagella. The roach supplies its partners with macerated wood, which it cannot begin to digest itself. The protozoans cannot get wood for themselves, but they can simplify its cellulose and produce substances that will nourish both protozoan and roach. Working together, the partners literally liquefy a fallen tree, and get the energy they need for life from the process. Separated, either partner starves.

Most termites are wood-eaters too, and possess some of the identical partners. Like the wood-eating roaches, they develop a critical need for companions every time they molt. Like any insect that sheds its skin, freeing its body to expand to the next larger size, the wood-eater also disposes of its gut contents, along with the delicate cuticle that lines its digestive tract. The stock of live protozoans must be replenished because, without them, the wood-eater can get nothing but constipation from its customary diet. How could restoration of this vital resource be more direct than by begging a droplet of gut contents from another roach or termite? Wood-eaters of both types of insects greet others in their separate colonies by sharing droplets, keeping every digestive system in good working order.

No one feels special fears for the safety of possessions after finding some wood-eating cockroaches in rotting wood. These most termitelike among the roaches pursue their secretive ways without endangering any human artifact. The most roachlike of termites earn a totally different reputation. They are the most destructive members of their family in Australia, and the most devastating insects of any kind in the northern half of that great continent. Each colony contains many millions of individuals. They attack dead wood, live wood, and crop plants, including vegetables and hay. They will devour horn, wool, leather, bone, ivory, rubber, plastics, and excrement. In two or three years, these termites in the Australian outback can transform an unattended homestead to a pile of dust.

"Termites have no business being so successful!" declared veteran entomology professor C. H. Kennedy. He told us that members of this order of social insects continue an ancient crudeness in their behavior. "Both males and females act as nursemaids in a termite nest, you know. All ages, from the very young to the fully grown, share in these duties. Much of the work force is juvenile. Child labor—that's what it is! And every time a young termite molts, it forgets most of what it has learned to do. It has to discover its duties all over again, getting guidance as well as food from other nestmates. How can any society succeed in so primitive a pattern?"

Actually, just about five hundred different kinds of termites —a fourth of the total diversity in this small order—rely extensively on immature helpers. They constitute about 50 percent of the total work force, and growth of the colony is slow. Eventually, however, the amount of queen substance no longer suffices to inhibit all young female termites from developing functional reproductive organs. In parts of the subterranean nest remote from the original royal pair, a few wingless virgin queens mature. Simultaneously the inhibitory secretion from the old queen's consort (the "king") fails to prevent every young male termite from achieving sexual competency. Yet the workers maintain close control. They bite and kill all but a few of the new queens and consorts. Far from others, sister and brother mate. She lays fertile eggs, adding young to the total population of the colony. No doubt these secondary reproductive individuals contribute also to the chemical control that regulates the expansion of the colony and keeps it adjusted to the resources it has available.

The original queen and her mate remain at the very heart of the nest. Workers guard them as their most precious resource. The queen alone can keep the colony coordinated. Incredibly, she detects in the food the workers bring her some subtle cues that keep her synchronized with the outside world —a world she will never see again. It is as though her liquid food both energized and continually adjusted a hidden computer in her body. She follows a natural calendar that will take full advantage of spring rains on the land around the nest. She will respond in season by producing eggs that are destined for a different pattern of development. The young that hatch will mature with good eyes and four flat narrow wings of equal length, rather than as wingless termites that are almost blind. A dozen or more will be females, each a queen ready for a suitor. Better than a hundred will be winged males of slightly smaller size.

The workers develop a special sensitivity when the winged assembly is ready. After dark or in the hours soon after dawn, according to the species, they test for a faint fragrance in the air. It will tell them if this is the particular night or day for the adults to swarm in synchrony with other colonies in the region. At the right hour, workers expose a dozen crescentic slits around the base of the termite mound, or open a few small portholes near the top of the structure. The males rush out of these special exits. On short legs they scamper, their long wings still folded over their backs but shimmering in any moonlight or the early sun. For a few minutes, the whole nest mound may seem silvered, glittering. Then the males fly off. Any that settle close to the mound are quickly cannibalized. Most males travel half a mile before stopping. Soon the whole countryside is speckled with the dark bodies of these insects from the many colonies. Each male waits, alert for the scent of a virgin queen of his particular kind.

Back at the nest, workers now shove the winged females through the doorways. When the last one is out, the mound will be sealed up. This is the only day for perhaps a year that the termites risk letting dry air enter their dark, humid passageways. The young queens are on their own. To be successful (and few are), each one must elude insect-eaters of many kinds. She flies a short distance, then runs over the ground in search of a place to found a colony. On soft soil, she pauses to flail her wings, dispersing a lure that quickly brings a male.

He follows her closely as she runs about, testing for a place to dig. Before actual excavation gets under way, both shed their wings as simply as though shrugging their shoulders. Wingless henceforth, they are queen and king, ready to start a family.

This could be the founding of a social organization capable of enduring for fifteen to twenty-five years, with the same queen as its central figure. Her body will continue to grow. Some get to be 100 millimeters (four inches) long, and the thickness of a man's thumb, no more mobile than a half-inflated blimp. Dozens of workers attend her. They carry off the eggs she lays, eight thousand to thirty thousand daily—one every few seconds—to pile in special chambers. Her consort stays close by, and refills her sperm reservoirs every week or so. He may need fifteen seconds just to run from one end of his mate to the other. Yet his lifespan rarely equals hers. Even though his contribution is microscopic, his virility falters or he dies from advancing age. His king substance will, by then, have failed to suppress maturation of new males. One of them, perhaps the worker's choice, will move in with the queen and keep her eggs fertile.

So long as the termite colony has its long-lived queen, it responds to messages in chemical form, ensuring its own survival. It wastes no energy or food on extra reproductive individuals of either sex until a particular type is needed to perpetuate the social group, or until both types can expand its heritage in extra colonies. The Belgian poet and author Maurice Maeterlinck recognized enough of this behavior for him to describe the colony as a single living entity. In his rather mystical account, *The Life of the Termites* (1926), he noted that the individual insects were as mortal as the individual cells in a person, but the community of them carried on. Probably Maeterlinck was unaware that Harvard University's outstanding entomologist, William Morton Wheeler, had applied the word "organism" to a colony of social insects, and offered in 1911 a definition that would support this view: "An organism is a complex, definitely coordinated and therefore individualized system of activities, which are primarily directed to obtaining and assimilating substances from an environment, to producing other similar systems, known as offspring, and to protecting the system itself and usually also its offspring from disturbances emanating from the environment." Like "un seul être vivant," as Maeterlinck described it, the colony as an organism centers its behavior around nutrition, reproduction, and protection.

Termites need protection from dry air and from raiding ants. They can combat dry air by sealing themselves inside interconnecting galleries, often over or surrounding items of food, before the material is cut to pieces and carried home. Raiding ants require a defense akin to a police force or a national guard. Even the first brood of a young queen will include a few sterile adults known as soldiers if they possess large powerful jaws, and nasutes if the armament is chemical. A nasute has a huge gland and nozzle from which acid or gas can be sprayed on an attacker. Any termite can summon help by emitting a volatile alarm substance. The nearest soldiers respond by rushing to the scene and biting at anything that moves if it lacks the termite odor. If the colony has nasutes too, they hurry to the center of disturbance and spray their repellent on any creature that bites them.

The defensive castes among termites are expendable, and commonly give their lives in safeguarding the colony. Once order is restored and the nest is securely sealed again, the surviving termites can assess how many defenders were lost by the diminution in soldier or nasute substance diffusing through the passageways. A lowered concentration of the volatile material is soon followed by a restoration of the guardian force. Once the normal complement is reached, these chemical messengers will inhibit maturation of more defenders, but an excess will induce workers to kill and eat soldiers or nasutes, to stabilize some suitable number.

A similar but separate set of messenger substances emanates from the workers. These chemicals help adjust the size of the labor force in relation to the food supply. When understaffed, the workers encourage the queen to increase her production. If too many workers are already present, those tending the queen will eat some of her eggs to prevent overpopulation.

We discussed some of these habits among termites with S. H. Skaife while he showed us a colony of one species he kept under close observation in his living room at Hout Bay, near Cape Town in South Africa. "Of course they engage in cannibalism, eating excess brothers and sisters, when this action will aid the social economy. And the secondary reproductives, male and female, in remote corners of the nest engage in incest to contribute more fertilized eggs for the good of the local population. But do the termites let anything get out of hand? The whole society maintains its health. I can never tire of learning how they do it!"

To us it seemed incredible that the black-mound termites of Africa, which construct such towering edifices outdoors from earth and feces, should adapt their complex lives so readily to an indoor world on a metal-sheathed table top. Yet workers regularly emerged to harvest the dry grain and grasses Dr. Skaife set out for them. Every week the pattern of their roofed-over runways changed as though to reach some new resources. Every year on schedule the termites had winged reproductives ready, and hurried them forth by day, as though the Skaife household had other locations to be colonized. Otherwise these insects remain, as our host entitled his delightful book on their behavior, *Dwellers in Darkness*.

Certain of the African termites, all members of the most advanced family (Termitidae), inherit from their long-lived queen and her consort a different specialty in diet. They are fungus-eaters and fungus-growers, hence agriculturalists. Between seasons of frenzied nuptial flights, the workers emerge nightly from the nest mound to get bits of grass and other foliage to carry home. Well chewed, the material goes into special galleries. There it is modified by the peculiar fungus. It alone has all the enzymes needed to convert the cellulose and lignin of plant cells into nourishing compounds. The altered plant material takes on the consistency of cork, forming pendant, curtainlike combs. Soft white buttons of fungus grow out from the surface. Termites devour the buttons and also bits of the modified plant material. This diet allows them to thrive with no need for intestinal protozoans. Perhaps the fungus is less vulnerable than single-celled animals to the high temperature that the termites tolerate within their mound.

Each virgin queen carries a dowry of the living fungus when she flies forth to found a colony. She conceals it in a special pocket. Tradition requires her to fill that pocket at the last moment before her nuptial flight, like putting on the wedding gown. It is her essential possession, an endowment to protect and never eat, until after her first small brood of sons and daughters grows old enough to care for it. The young queen keeps the fungus alive with droplets of excrement. She feeds her young with tiny eggs, sharing her own meager resources. These eggs contain nourishment derived from the internal digestion of her wing muscles—organs she will never need again. The queen has nothing else to offer, for she has had nothing at all to eat since she left her parental colony. None too soon, her brood matures and begins foraging for foliage to feed the fungus, to produce food for the fasting queen and king, then for an ever-growing population of their particular colony.

Each nestful of fungus-growing termites recycles bits of fallen vegetation far faster than would occur without their work. A large colony may take in and process half a ton of plant materials annually. But when the annual rains begin, the insects start housecleaning. They dismantle their fungus garden and spread every crumb of it on the soil surface outside, close to the mound. The fungus strands join up and raise a whole crop of mushrooms into air, while the termites sanitize their subterranean chambers. Most fungus spores ride the breeze for great distances. Some settle to the ground. The termites rake up the soil from beneath the mushrooms and haul it to the renovated chambers. There the fungus spores germinate on a new supply of chewed vegetation, like a seed crop the insects sow to carry on their way of life.

Only Africa has fungus-growing termites, and only the New World has fungus-growing ants. The parallels impress anyone who recognizes how separate paths of evolution reward unrelated insects for such similar patterns of social behavior. Yet, as Dr. Skaife insisted, the termites alone inherit programs of action we can admire and compare with the best among the cultural systems of humankind. Unlike the ants, these supersocial insects engage in no destructive wars. They neither capture nor maintain slaves, and never become parasites. Termite societies accord virtually equal rights and responsibilities to members of both sexes. Silently, with no commotion, they attend to succession when their royal members age or falter. And if a soldier loses its head after biting an adversary, it sacrifices its life without aggression, without poison, and generally without a sound.

Charles Darwin explained the sacrifice of nonreproductive social insects in *The Origin of Species:* ". . . we may safely conclude from the analogy of ordinary variations that each successive, slight, profitable modification did not probably at first appear in all the individual neuters in the same nest, but in a few alone; and that by the long-continued selection of the fertile parents which produced most neuters with the profitable modification, all the neuters ultimately came to have the desired character."

Darwin's explanation of sterile castes and altruistic behavior satisfied biologists for more than a century. It did not show why termite colonies are organized so differently from those of social members of the order Hymenoptera (the ants, wasps, and bees), as they too associate with a long-lived queen. Nor did anyone understand why a colony maintains few of its most efficient castes. Why not more, if individuals of the caste prove better adapted?

Social behavior seems a logical development in insects that can exploit dry wood for food because they possess suitable intestinal partners. This lucky symbiosis, progressively exploited in both sexes, would largely account for the evolution of the whole order of termites (the Isoptera). Yet social termites account for only about two thousand species. By contrast, the social ants, bees, and wasps total close to fifty thousand living kinds—just over eight percent of all the insect kinds in the world. How did social hymenopterans come to diversify so much more than the vast majority of animals, and to elevate the various roles of females to such a special level?

An English biologist, W. D. Hamilton of the Galton Laboratory in the University College of London, offered a reason in 1964. His "genetical theory of social behaviour" is still regarded everywhere as audacious. Hamilton suggests that sociality could scarcely be avoided among female members of order Hymenoptera because of their peculiar mode of reproduction. As has been known for many years, sterile workers and queens are females that develop from fertilized eggs. Males in this insect order come from unfertilized eggs. The difference biases the benefits of parental efforts in care of young as though the laws of chance had been distorted with loaded dice.

Hamilton's view goes straight to the ultimate molecules in which the heritage of each species is carried in coded form— the genes. They specify how the individual can develop and what actions are possible, and they perpetuate those combinations that can be successful in the environment. The insect (or any other organism) becomes the means whereby a winning combination of genes perpetuates and propagates their detailed specifications. This is a step beyond the old claim that a hen is an egg's way to make more eggs.

The behavior of each insect then holds importance only to the degree that it enhances the proportion in the next generation that possess the same genes. Hamilton could predict this in terms of kinship, which is so peculiar among the hymenopterans. Each female worker is a female because she had a father; her double set of genes includes one set that came from him. An identical set went to each of her sisters, making one half of their inheritance alike—that from their father's side. But each female received also one or the other set from the two her mother possessed. Two sisters either receive the same set of genes from their mother, and are one-half alike on their mother's side too; or they receive unlike sets of genes from her, and are unrelated on the mother's side. A sister shares with other sisters one quarter of their inheritance alike from their mother, plus one half of their inheritance alike from their father. A sister shares with other sisters three quarters of the parents' genes. She has one half of the genes carried by her mother, and can be represented only by one half among the genes of any offspring she herself may have. She shares only one quarter of the genes inherited by her brothers, since she either received the same set that the male did from her mother, or received the other set and is no kin to him at all.

Hamilton argued that a female hymenopteran contributes more to her species by safeguarding the female young of a prolific sister—the queen—than by having offspring of her own. She should allow little time or food for a brother, since her chance of sharing genes with him (and he of carrying on her heritage) is only one quarter. True, her species needs him to fertilize a queen at swarming time. Why have males around at all at other seasons?

The queen in any colony of hymenopterans provides as many of her genes to a son as to a daughter. She might be expected to produce equal numbers of unfertilized and fertilized eggs on the basis of chance. But can the workers that care for the eggs and young recognize the sex of each? They might prevent males from maturing toward their destiny. This seems to be true early in the queen's life. As the queen ages, more males are produced. Is there still a tendency to favor virgin queens, rather than drones, in readying offspring for the swarming season?

Any tendency that benefits the species in nests with a long-lived queen or two would be enough, if it is real. The lowest rate of interest yields a spectacular increase if the compounding occurs at each generation for millions of years. Time is the wild card in the game of life, the "hero of the plot," in the words of Harvard University's George Wald. "Given so much time," he told us, "the 'impossible' becomes possible, the possible probable, and the probable virtually certain. One has only to wait; time itself performs the miracles."

Mating season is the only opportunity for a male hymenopteran to make a contribution to his species. Each of his sperm carries a single set of genes. Yet whether it will fertilize an egg or not, and whether the larva the egg produces will be permitted to develop, is under the females' control, not his. He can produce no sons. His daughters all receive the same genes from him. He may as well be a lazy drone, taking every opportunity to seize a mate (even if completion of the act ends his short life) because his social sisters will limit the free food he gets.

If the workers have been democratic in distributing nourishment to the brood that will become virgin queens and drones at mating season, the total weight of young queens and drones should be approximately equal. If, on the other hand, the workers consistently give out food three times as generously to female larvae (their younger sisters) as to male larvae (their younger brothers), then young queens should outweigh the drones threefold. Robert L. Trivers and Hope Hare at Harvard University decided to compare these weights as a mea-

sure of what actually occurred in the nests of twenty-one different kinds of ants. The average came to 3.36 times as much dry
weight in virgin queens as in drones, showing that the workers
biased the outcome beyond the proportion predicted from
Hamilton's kinship theory.

Some ants spend no energy in raising offspring of their own
species. The workers, instead, raid colonies of other ants, kill
the workers, carry home the helpless pupae, and allow captive
workers to emerge. The captives are slaves, with no queen of
their own and no chance to reproduce. They tend the eggs and
young of the slave-makers' queen. In their own colonies, the
slave workers would bias the distribution of food to raise virgin
queens weighing three times as much as the drones. As slaves
they perform no such service. The slave-making colony disperses an equal weight of new queens and drones, showing
that the slave-makers have lost control over the frugal distribution of resources toward their own reproduction by relinquish-
ing all care of the brood.

The genetic heritage alone guides the social insects. They
let the statistics of survival adjudicate the outcome. Yet their
genes also call forth complex behavior, which regulates the
proportion of each caste within the total population. No one
has discovered how the colony maintains these differences in
weight and numbers, but the advantage in doing so can be
recognized, as Edward O. Wilson of Harvard University has
done.

These extensions of Hamilton's kinship theory help us understand the probable evolution of social insects. New questions can be asked, or old ones raised so clearly that tests can
be devised. If defense is the prime task for soldiers, how much
of an army must the colony maintain? Surely the aim is to avert
any serious attack, to avoid a significant reduction in the number of virgin queens that can be dispersed at mating season.
How often will the soldiers prove essential? How many will be
needed to cope with sudden needs? Similarly, how many workers and in what sizes are needed to get food and to care for
the brood? The actual numbers in a colony can be counted and
used as a measure of the contingency.

Each sterile worker represents an investment of food from
the colony's resources, and special services must earn it back.

These are tasks that the queen cannot manage without reducing her output of eggs. Wilson notices that the more different roles each worker can perform, the more of such individuals the colony will include. Large soldiers, each powerful and efficient, will be few—a proof of their superior adaptive value and cost. Small soldiers, which must act in groups to protect the colony, will be numerous. The queen herself is the ultimate specialist. So long as she produces eggs at top speed, the colony needs only one of her kind.

The environment challenges each insect colony to maintain an optimum mix of its various castes. Economy dictates "Nothing in excess," just as the Roman writer Terence recommended in the second century B.C. Food must not be diverted from the reproductive line to no lasting advantage. This explains why the more effective a sterile member is, the fewer of this caste the colony will support on its limited resources.

A PROGRESSION OF BEHAVIORS

The domesticated honeybee reveals an alternate pattern in the division of labor among sterile workers. Rather than have a separate caste to perform each task and to meet various contingencies, the heritage calls for workers of one kind. They specialize on a particular task (or a limited number) for a while, then progress to different duties. The body of the worker and her inherited behavior need not at any moment be a complete compromise in the direction of universal competence. Her physiological changes can be progressive, and permit the sequence of roles she must perform. Her behavior shifts every few days, or oftener.

The worker honeybee acts as though she had a detailed work sheet to follow each day from her emergence from the pupal cell until death overtakes her. Her first labor is to force open the waxen doorway of the chamber in which she transformed from grublike larva to adult. Next she must clean herself all over—a task that workers did thoroughly for her during each of her larval stages. Now, in the darkness of the hive, she encounters older workers whose crops are filled with nectar. She begs droplets from them as her first adult meal. After cleaning herself again, she crawls to the new brood comb that other workers have just completed. She enters one empty chamber after another and swabs the inner surface of each with her saliva. Until this is done thoroughly, the queen will not accept the cell as a suitable place in which to lay an egg. The young worker stands aside as the queen and her retinue move about. In fact, the first three days pass while the young worker does little else, except to stand on the surface of sealed cells of brood comb. Perhaps she contributes a small amount of body heat to the pupae that are developing inside.

After her third day, the worker visits the honeycomb and takes food for herself directly. She eats more pollen at that time than at any other in her life. Apparently it stimulates the development of special glands in her pharyngeal region. From them she will later regurgitate a particularly nutritious "brood food" for the youngest larvae. But on Days Three to Five of her adult life, she has only honey to offer. This she takes to the oldest larvae, then to progressively younger ones. By Day Six she is feeding the hatchlings too. She continues servicing the young in the brood comb until she is thirteen to fourteen days into adulthood. Now her pharyngeal glands are shrinking fast. Below her abdomen quite different glands are producing wax. Between Days Twelve and Eighteen, she uses that wax to fashion new cells of the standard hexagonal pattern in regions of new comb—honeycomb and broodcomb.

Day Twelve or thereabouts brings a special event. For the first time the worker ventures to the doorway of the hive. She backs to the edge of the sill and expels a white cloud of dust. It is composed of uric acid crystals—her nitrogenous waste—along with a small amount of fecal material. She has held back all of this for three weeks of larval and pupal life, plus almost two weeks more while moving about over the comb. Now, and at least daily hereafter, she relieves herself outside the hive.

If the weather is fine, the worker may learn some of the local landmarks during a few brief "play flights." If the day is hot, she stands instead on the sill and holds firmly, while using her wings to fan outside air into the hive. Or she waits just inside and accepts the loads of nectar, pollen, water, or sticky resins (called propolis) that older sisters ("field bees") bring one after another. She frees them to fly on further missions. The nectar goes into the open honeycomb to be cured and considerably dehydrated. The pollen she packs into special cells, and the resins wherever any crevice leaks air or light. Water is spread around to evaporate and cool the hive, as well as to sustain its moderate humidity.

By Day Fourteen, the worker stands on the sill of the hive many hours daily, checking field bees as they return, repelling any that lack the proper hive odor. Each field bee to touch down and crawl through the doorway opens a special pocket where this odor lingers, as though she had a pass badge to display. Or, at two weeks of age, the worker polices the inside of the hive to pick up any fragments of wax, any pupal skin, or similar debris. She hauls these to the door and dumps them to the ground. If she discovers another worker that has died inside the hive, she gets help in disposing of the body in the same way.

The third week of a worker's adult life may be her last. Her heritage specifies that she should act as a field bee until her wings wear out and her nonreproductive body ceases to contribute to the queen's output of young. She may concentrate on collecting pollen for the early part of this final week, and nectar later on. She will respond to subtle signals from workers at the doorway that inform her of special needs in the hive —for water or resins in particular.

A field bee learns quickly which colors and patterns of flowers yield the resources she seeks, and what hours of the day they are open for entry. All of this she will remember, as well as her way home by the straightest route—a "beeline." Less of the energy obtained from honey need be expended without benefit to the colony if she heeds her inner sense of passing time by resting while the flowers are closed, and by navigating expertly with her "sky compass" between the hive and nectar sources.

The worker can communicate some of her discoveries to others of her age group inside the hive. Karl von Frisch identifies the message in terms of peculiar waggle dances performed amid a group of attentive workers on the vertical surface of the comb. He believes that direction, with reference to the position of the sun, and approximate distance, with the uncertainty no more than 20 percent nearer or farther, are the primary pieces of information. Scent from the nectar brought home by the worker may serve as additional guidance. Perhaps this is true for the honeybees von Frisch studies in Europe. Elsewhere, scent may be more significant. Experiments by A. M. Wenner and his students in California support this alternative explanation. Yet, regardless of the nature of the message, the worker does stimulate others in the hive to benefit from her experience. All too soon, everything she learns dies with her.

As long as the field bee lives, she stays ready to protect her queen. A whiff of an olfactory message (an alarm pheromone) from another bee transforms her and her sisters almost instantly into a suicide squadron. They seek a target to attack. Actually, this chemical call to sting has a dual origin: one component from a gland close to the stinger shaft, released as each bee stings; and another from mandibular glands, which alarmed bees expel to recruit assistance. The honeybees of the African race apparently produce the mandibular-gland secretion more copiously than Eurasian strains, and charge upon any living thing in their path when the colony is disturbed. Too often, this makes them "killer bees."

Otherwise the African honeybees behave in ways that apiculturalists admire. The workers are busy earlier in the morning than Eurasian bees, work later into the evening, and venture forth on cool and rainy days. They make more honey. Yet they respond to cold weather in the same way as the gentler strains: they stay in the hive and cluster around the queen, vibrating their muscles to convert the energy from honey into heat and keep the royal body warm. This is the only activity that prolongs a worker's life—for several months, if winter lasts that long.

A friend who keeps a dozen or more hives of honeybees tells us how many workers behave on the first warm day of spring. As the sun raises the temperature a few degrees inside the wooden box, these bees leave the cluster around the queen. They creep out on the doorsill, and to its edge. "Then they die. Just like that!" Apparently they hold back just enough energy to make this final trip, so that their corpses do not clutter up the inside of the hive.

Everything possible is done to help the queen stay healthy and productive for five years or more. She too inherits a set of special directions to follow in her first week of adult life:

hunt out and sting to death every other virgin queen maturing inside the hive (strangely, a queen can sting and survive, whereas a worker cannot); lead the drones on a nuptial flight or two; accept every sperm from one suitor after another, before casting him aside to die; and when storage organs are filled with a lifetime supply of sperm, return to the hive to assume the royal role. Months later, if the quarters become cramped with a large reserve of food and brood, the queen will follow the scout workers amid a swarm of field bees to a new location. Enough workers will remain behind to serve a new queen as she begins the same sequence after a day or two. In each colony, the queen's behavior and odorous messages, more than those of any other individual in the hive, seem to control the sequence of events.

A one-inch honeybee or a smaller ant stimulates us to "think small" better than any other animal we know. Like a marvel of miniaturization, each offers proof that a limited array of components (cells) can provide for an immense range of tasks and contingencies, so long as those components are of the correct kinds and superbly organized. So small an individual can specialize and be efficient in role after role, even during a three-week lifetime as an adult, following an equally brief period of juvenile development.

THE SOCIAL ARCHITECTS

An anonymous monk of the twelfth century wrote the following in a scrapbook that is preserved in the Cambridge University Library. T. H. White translated it from classical Latin, for the benefit of readers in modern times:

> What indeed is a honey hive except a sort of camp? For these enclosures the bee-wax of the bees is laid up. What four-walled houses can show so much skill and beauty as the frame-work of their combs shows, in which small round apartments are supported by sticking one to the other? What architect taught them to fit together six-sided chambers with their sides indistinguishably equal? To suspend thin wax cells inside the walls of their tenements? To compress honey-dew and make the flower-granaries to swell with a kind of nectar?

The array of cells we know as honeycomb, if it serves for storage, and as broodcomb, if it contains immature bees, is certainly the best known and most widely respected product of insect architecture. The pattern of open or closed cells is as perfectly hexagonal as fine Brussels lace. It has excited the mathematically minded since the dawn of science. Pappus the Alexandrine suggested in the third century A.D. that honeybees economize on wax by building in this way, and thereby prove themselves endowed with "a certain geometrical forethought." The crystallographer Erasmus Bartholin in the seventeenth century may have come closer to the truth. He judged that the pattern arose because each honeybee strives to make a new cell as large as possible, and squeezes against others in the construction process.

No honeybee helps another in this enterprise, or waits to be brought building material, or is called upon to hold a piece. Each bee brings her own wax. She arrives to use it only when the wax is ready, and departs as soon as her day's supply runs out. If this leaves a cell unfinished, a fresh worker with a full load of wax steps in and completes the job. No task is a team effort, although a dozen or more bees labor side by side simultaneously. The cells of the comb seem mass-produced, despite the number of independent bees from the colony that participate. Each insect uses her own body as a measuring device while forming the wax. The dimensions stay uniform because all workers are so similar in size.

No bunk bed for people would be acceptable if it were as confining as each cell in a honeybee's nest. This standardized space is only slightly larger than the insect's body. Yet its general shape, with its long axis horizontal, lets gravity prevent the contents from falling out. (Wasps and hornets make vertical cells, which a wasp enters from the top, a hornet from the bottom.) Every cell in the compact tier that honeybees make shares a vertical wall on each side with adjacent cells. The low gable roof is floor material of two cells above. The floor of each cell seems provided for drainage down the middle. But no reinforcing strands or insulation thicken the wax. It is paper thin.

In proportion to the bee's body, the topmost cell may be ten body lengths above the lowest cell. A person would need to be a veritable human fly to clamber sixty feet above the ground without a ladder or a rope, all to reach a chamber the size of a minimum coffin. Yet the bees not only crawl up and down without hesitation; they also perform special dances on the vertical surface, as returning workers communicate to others the distance and direction to a good source of food.

As many as fifty thousand adult honeybees may shelter in the same hive every night during part of the year. They are all brothers and sisters except for one—the queen mother. Accompanying the fertile queen are perhaps three thousand of her full-grown sons (drones). The remainder are sterile daughters, who do everything except lay the eggs. Uncounted young will be growing in the brood comb. They, like the stores of honey and pollen, cannot help but be under the feet of the adult insects taking refuge in the hive.

The worker bee that brings food to larvae in the brood cells has no help at all. If she has reached the age at which her own glands are secreting the special food the youngest larvae need, she alternates between attending to them and seeking food from the storage reserves inside the hive, to keep herself well nourished. Earlier and later in her adult life, when she is waiting on older larvae, she dilutes the honey she withdraws from the honeycomb with her own saliva before regurgitating the liquid to the brood. The insect acts as a surrogate for the queen, but not as a waitress or a nurse bee, although these roles are frequently ascribed to her.

A different contribution, made by every honeybee inside the hive, is that of body heat. Each honeybee generates about one-tenth of a calorie per minute. A human body, according to the calculations of air-conditioning engineers, emits about twelve hundred calories per minute—about the same amount of heat as a 100-watt light bulb. A hiveful of bees can produce as much heat as four people, crammed into less space than any full-grown person would require. This explains how the uninsulated hive is warmed, but not how the insects keep the temperature close to 92° F. (33° C.) for about ten months of the year, while larvae are developing in the brood cells.

Maintaining an imitation summer inside the hive for so many months hastens the digestive and growth processes of the immature bees, both larvae and pupae. It hurries the evaporation of water from the honey that is "ripening" in the open cells of honeycomb. Progressively that water content diminishes from around 90 percent to no more than 19 percent. Less than 18.6 percent water is the requirement of the U.S. Department of Agriculture for Grade A honey. This amount is too little to allow yeasts or bacteria to grow on the stored food, and it keeps well—particularly in the sealed cells of the comb. Fermentation during the ripening process seems inhibited by secretions added by the bees, both while they are sipping nectar from flowers and later, in the hive, as they tend the honeycomb.

A trade-off occurs when the bees accept an artificial hive as a nest, instead of a hollow in a tree. The old tree ordinarily provides thicker insulation against winter cold, and its hollow trunk is shadier. Space for more comb, and sheets of waxy foundation, are the lures the beekeeper offers and the insects accept.

A dark cavity seems a normal requirement before honeybees will begin to build their waxen cells. The senses of scent and touch suffice. No windows or extra doors are needed. The bees leave few openings between their comb and the outside world, through which unwanted heat might enter or their own body heat escape. Any space more than a centime-

Compass termite nests in northern Australia, often taller than a man, are oriented to minimize heating by the sun at noon and to maximize warming by the sun early and late in the day. (Photo courtesy Australian News & Information Bureau)

ter (three-eighths of an inch) broad is soon filled with extra comb. Every crack too narrow to admit a bee gets filled tight with sticky resin (propolis), which hardens as the insects' substitute for caulking. It limits air flow and effectively shuts out invaders too.

Inside the nest the bees tolerate a temperature that a person would regard as stifling. They feel no need to cool the hive until heat from outdoor air, direct sun, and the living occupants themselves raises the temperature above 35° C. (95° F.). These occasions come less often than might be anticipated, if only because so large a proportion of the worker population is out of the hive during the hottest hours on foraging expeditions. Nor do they return while the sun is high, even though the flowers yield so little nectar at that time. The field bees tend to settle somewhere in the shade and rest. Inactivity diminishes their heat production, and the energy is dissipated far from the brood.

The resident workers, as contrasted with the field bees, are mostly younger adults. They too adjust their behavior on hot sunny days. Many bees move to the doorsill and to locations on inner surfaces of comb where they can fan their wings. Efficiently, although acting as individuals, they propel outdoor air into and through the nest. An open space lets the hot air escape. When cold weather returns, all gaps are closed tightly. Even the front entrance is constricted until it is only about one bee wide.

The honeybees abandon most enterprises when the weather cools to or below the freezing point. By then they should have a reserve of honey and pollen for winter use. If the population of the colony weighs four pounds, the bees will need approximately an equal weight of carbohydrates to see them through the winter. They will not attempt to keep the food store warm, but instead cluster in groups, particularly around the queen. This behavior distinguishes the members of the genus *Apis* from bumblebees, other bees, and wasps. All through the cold weeks the honeybees will keep their queen at 20° C. (68° F.) or above, and trade places in the cluster both to go individually for food and to let no bee on the surface of the cluster get too chilled to move. The usual minimum is probably not much below 10° C. (50° F.). Once more the insects of the honeybee hive demonstrate an inexpensive way to achieve environmental control by localizing its application.

The queen bumblebee employs a different strategy. Like many of the queen wasps, she hibernates in some protected place. But unlike the wasps and the honeybees, she emerges early in the growing season. Clad in an overcoat of golden hairs, she collects nectar and pollen to fill little honeypots of her own construction. She makes a few brood cells of smaller size in a cluster, and drapes her plump body over them, facing a honeypot so closely that she can reach a sugary snack by extending her mouthparts. All night and on any cool day, she shivers steadily, thereby keeping her thorax and abdomen at a temperature above 31° C. (88° F.). Like an electric hot pad, she keeps her eggs and young in the range between 24° and 34° C. (75° to 95° F.), even when the surrounding air and the food in her honeypots are no more than 3.3° C. (38° F.). Every warm day she leaves her brood at intervals to fly off and gather food, refilling her reserve. She will continue this routine until her first daughters hatch and help. By the time summer arrives, the aging parent and several adult offspring—usually a size smaller—will tend the informal colony. Often they have no more shelter than a small pocket in the cool earth. It may be a simple nest in the side of an entry tunnel to a chipmunk's winter hideaway.

Insects began taking advantage of the insulating effect of the earth itself long before humankind appeared. Bumblebees, ants, and termites employ methods that engineers and architects may soon feel impelled to imitate in sheltering people against summer heat and winter cold with a minimum expenditure of energy. The sod house and the yurt, half hidden by surrounding soil, offer benefits that merit exploitation. The mounds of the compass termite near Darwin in arid parts of Arnhem Land, Australia, are especially conspicuous from a small airplane flying east or west at low elevation. Each mound is 3.0 to 3.7 meters (10 to 12 feet) high, barely more than a meter (3½ feet) thick, and 3 meters wide in a north-south line. Morning and evening it catches the low sun and soaks up heat. But near noon it casts little shadow and minimizes the need for cooling operations inside.

Martin Lüscher of the Zoological Institute at the University

A swarm of honeybees sometimes cannot find a suitable place to harbor their queen, and prepares pendant vanes of comb below a large horizontal branch on a big tree (here a sugar maple) in deep shade. When the foliage fell in the autumn, the nest was exposed; all the bees would have frozen to death, had not a beekeeper rescued them and given them a weatherproof hive for the winter and future.

of Bern in Switzerland has told us details of the inner organization of bulkier termite mounds he has investigated in East Africa. The tallest and most spectacular—to almost five meters (sixteen feet) high and equally broad in all directions at the base—are constructed by *Macrotermes natalensis.* It is a wide-ranging insect, which builds in forest and open areas alike, generally where the soil is an orange-red laterite. It mixes mineral particles with saliva to make a hard cement, and produces a community dwelling with a firm foundation. The walls are forty to fifty-eight centimeters (sixteen to twenty-three inches) thick, and a conical top sheds all but the most persistent rain.

The imposing fortress surrounds a maze of interconnected small chambers, a queen compartment that would accommodate a liter (quart) can of motor oil, and a vacant attic where hot air can accumulate harmlessly. The inner architecture is all free form, graceful, based on the arch rather than symmetrical reduplication like the combs of the honeybee. The termites create surprisingly intricate channels for air flow and drainage, and change them unpredictably.

Lüscher drills holes through the walls and installs electric thermometers to learn how well the termites air-condition their huge nest. He knows that hot air from the attic moves through major ducts to fine channels, just beneath the outer surface of external ridges on the sloping sides of the mound. As the air moves close to the surface of the nest, diffusion occurs through microscopic pores in the brick-hard material. Carbon dioxide from insect respiration, and from the fermentative work of fungi in the central compost heaps, is replaced by oxygen. Heat escapes and is carried away by an outside breeze. Inside, the cooled air has greater density, and gravity causes it to descend. Thence it circulates inward and upward again, through the center of the nest, on another cycle. Lüscher waits until the termites seal in his thermometers. Then they record a change of 5° C. (9° F.) as the air moves from attic to basement through the cooling system. In such a termite mound, the air need not flow faster than a snail's pace to circulate ten times a day. No termite finds itself in a detectable draft. Fully 240 liters (roughly eighty gallons) of gas are exchanged daily, freshening the air inside the nest.

The elaborate air conditioning works well only after a termite nest reaches a critical size. The insects cannot control their environment suitably in small mounds such as those that a young queen and her consort begin when they found a colony. Lüscher believes this prevents the continent's most successful species from spreading northward across the great deserts into North Africa and Europe. The dry heat is too much for pioneers. "If evolution ever permits *Macrotermes* to form 'worker swarms' as bees do," he says, "or bud large complete colonies from the original mounds as some species of termites already do, no climatic barrier will be strong enough to hold it back."

The ridges of the giant termite mounds serve the same role

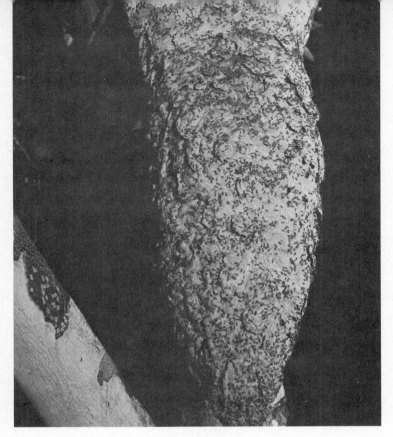

Many tropical ants build carton nests high in trees, such as this one in Panama, affixing the thin paperlike layers to some branch or the tree trunk. This construction serves as shelter for the queen, her eggs, and her brood, close to the hunting grounds of the ants, which almost never descend to the ground.

as the metal fins of high-efficiency radiators, and permit exchange of respiratory gases as well. Someday this design may appeal to an architect who decides to make the whole exterior of a building a functional part of the heating-cooling system. All the other structural features in termite nests seem to have been tried separately: the immensely thick walls of brick or stone around old houses, forts, and castles; the fungus garden (for edible mushrooms, at least) below ground level—the coolest part of the structure; the unused attic, where the hot air caused by a summer sun on the roof will be far from living quarters (our small New England home has this amenity); small rooms for the growing young and the live-in workers

who look after them; and a huge master bedroom for the royal
pair.

Some of the African termites provide their private nests with
a secret well. They dig downward vertically until they reach the
depth of ground water. They can go to this level whenever the
season's heat makes air cooling inadequate and threatens to
lower the humidity of the nest. The termites bring up water
and spread it on the walls of chambers and galleries, where it
evaporates and absorbs the excess calories.

The social insects show no need for water pipes or sewer
lines. The adult workers, who groom themselves and the
queen too if she is preoccupied with producing eggs, also lick
the larvae clean. Our nearest approach to this personal sanita-
tion without water or a drain comes when we use moist wash-
up tissue from foil or plastic envelopes. Nor do the insects
need training in disposal of wastes. Automatically they go
outdoors "for a walk," or tuck deodorized and sterilized pellets into the fungus garden as fertilizer.

GUESTS THAT COME TO STAY

In the fable of the grasshopper and the ant, Jean de la
Fontaine told about the grasshopper that stridulated a love
song all day, while an ant industriously gathered a store of
food for the coming winter. When cold weather arrived, the
grasshopper could find no grass to eat, and begged from
the ant in vain. In La Fontaine's day, neither the life history of
the common grasshopper nor what goes on in an ants' nest
had been examined closely. Until recently, no one realized that
in both Eurasia and North America, the smallest grasshopper-
like insects move right in with the ants and quietly help themselves to the ants' provisions. So do various cockroaches, bugs,
fly larvae, a few moth caterpillars, and numerous beetles. The
ants tolerate these "guests" most of the time. But any "guest"
must pass frequent inspection. It must be ready at a moment's
notice to appease any ant that shows hostility.

Occasionally we compare notes with Bert Hölldobler of
Harvard University, who takes special interest in all of the
insects that share the food in ant nests. The fact that so many
of the "guests" resemble ants in size and behavior impressed

Hölldobler as an adaptive mimicry. It confers freedom from attack by any insect-eater that would avoid an ant. Appearance is of little consequence when a guest meets an ant, either outside in daylight or in the darkness of the nest. "The question then," says Hölldobler, "is whether the 'guest' has the right odor and reactions."

At least the small wingless cockroach can follow the Texas leaf-cutting ants home, to partake of hospitality in the ants' subterranean fungus gardens. The roach keeps its antennae raised above the ground, but continuously taps against the soil a long pair of jointed palps from its lower lip. Like odor-sensitive fingers, they inform the cockroach when it finds a trail left by ants of the proper kind. The trail is no more than a succession of streaks of a chemical substance, which the ant applies to guide other ants (or itself) to and from the nest. Yet the roach seems as capable as the ant of finding the home-bound direction. This roach will also respond to the trail secretion from a different ant that commonly makes small nests superimposed upon the larger, deeper nests of the leaf-cutters. The roach shows a preference at the doorway of the nest, and associates with the leaf-cutters rather than with the different ants that live at penthouse level in the maze of passageways beneath the surface.

At certain times, the guest cockroach takes a more direct route. It holds firmly to the body of a virgin queen as she leaves the leaf-cutter colony to go off on her nuptial flight. Later, the mated queen returns to earth to prepare a new nest and found a colony. Her hitchhiking guest is still there with her, holding on, ready to move in as soon as she makes a suitable space.

Hölldobler also told us how one of the little rove beetles that lives with ants gains entry to the nest. The beetle finds the opening to the nest of the correct kind of ant by odor, but waits outside the entrance until discovered by a worker ant. At the first touch of the ant's sensitive antennae, the beetle turns quickly and offers the tip of its abdomen. There the beetle exudes a tranquilizer substance from a "peacemaker gland" in the cloacal area. The ant licks off the exuded droplet, whereupon the beetle turns and brings to the mouthparts of the ant a different set of glands along the sides of the beetle's abdomen. These glands release a different chemical message—an adoption signal—to the ant. A single sample seems to be enough. The ant reaches forward gently with its big jaws and picks up the beetle. The beetle curls up into a compact load and lets the ant carry it down into the nest. There the beetle soon acquires the nest odor—a distinctive fragrance combining the scent of the particular ants in all their stages of development with the scents from their store of food. If the nest odor, like a pass badge, does not satisfy any ant the beetle encounters underground, the beetle responds to any show of aggression by quickly repeating its messages: first the tranquilizer, then the adoption signal.

Rove beetles of various kinds get their nourishment from the ants in different ways. Some that we have watched will march straight up to an ant, particularly one that has just returned to the nest with food. The beetle uses its antennae to stroke the ant, soliciting the regurgitation of a nourishing droplet. The ant may not get the message. It may even spread its jaws as though ready to attack the beetle. Promptly the beetle brings from its own crop a droplet of food, like a small balloon of bubble gum. The beetle presses this droplet against the ant's mouth, and instantly changes the encounter from one of confrontation to one of collaboration. The ant regurgitates a generous blob of food, and the beetle drinks it down. This resembles the sequence by which an ant or ant larva gains food from a returning worker. It may be more successful than trying to raid the storehouse of dried foods the ants have stockpiled.

A beetle that begs is presumably a better guest than one that raids the brood chambers and eats ant eggs or young larvae, as some do. Unfortunately for ants, most of the beetle larvae that develop inside their nests are carnivores, living at the expense of their hosts by inconspicuously devouring eggs and brood. Yet, even in this activity, the beetle larvae limit their own depradations. Each larval beetle prefers to attack and devour another larval beetle, if two of them turn up in the same chamber of the ant nest.

A rove beetle that Hölldobler observed in Germany manages to be adopted into nests of two different kinds of ants. In spring, the beetles migrate to the forests, and there are accepted into the underground colonies of a particular forest ant

(Formica). All summer the forest ant brings in food and raises new generations of young, affording the beetle a supply of eggs and ant larvae. But in autumn, when the forest ant suspends these activities for the winter, the beetles emerge and travel to open fields. There an ant of open country, a *Myrica*, has extensive stores of high-energy foods ready for the use of maturing adults, which emerge on mating flights and found new colonies in early spring. The beetles and their young feast on the ants' provisions all winter, until longer days and warmer weather signal a return to the forests, to be readopted by the summer hosts.

The secretion used by the guests apparently closely resembles a substance that larval ants produce. Even a dead larva of the correct kind of rove beetle releases enough of the critical message from huge gland cells at the side of each body segment that an ant will pick up the corpse and carry it to the cluster of live ant brood. Experimenters find that the glands can be squeezed to smear the substance on short bits of twig or twists of paper. The ants will adopt these too, and treat them as though they were ant larvae. Surely this is olfactory mimicry used as a special kind of fraud.

Some rove beetles that cannot provide the adoption signal may still be able to secrete the tranquilizing substance. Some of these beetles enter the doorways of ant nests and accost returning worker ants where they are accustomed to transfer food to nest mates. The beetles sneak up under the mouthparts of two ants, and steal food during the mouth-to-mouth transfer. If either ant appears to detect the thief and interrupts the process, the beetle quickly offers a droplet from the peacemaker gland at its posterior end. As soon as the ant quiets down, the beetle hurries away, because tolerance is only temporary. A fresh approach to another returning ant is a safer way to prolong its meal.

Other rove beetles are the insect equivalent of muggers. They lurk beside ant trails until a well-loaded worker ant comes along, homebound. As the ant is about to pass, the beetle backs up for a deliberate collision and offers a droplet of tranquilizer. The ant stops, accepts the droplet, and acts distracted. That moment of apparent indecision is fatal, for the beetle leaps atop the back of the ant, decapitates her with a

single bite, and proceeds to devour the body, food load and
all.

It is tempting to arrange in a stepwise sequence the various
habits of rove beetles as they encounter ants. Most rove beetles, in our experience, are versatile predators, ready to
pounce on, kill, and eat any small edible creature they can
subdue. To choose ants like a highwayman or a common ban-
dit, killing and robbing, seems a minor specialization. To ac-
complish this in the doorway of the ant's home or, worse, in
the nursery, shows even greater refinement of technique. It
also reveals how few indicators the ants use in rejecting or
accepting another individual in their society. Termites and
their guests interact in much the same ways.

Scientists with a special interest in social insects and their
guests show greater unanimity in recognizing parasitic and
predatory actions than in deciding how the various patterns of
interaction evolved. Did the ancestors of certain guests discover that they, fortuitously and in advance, possessed the
chemical keys to the food stores, and then move in to help
themselves? A behaviorist would call this "resource tracking,"
and see a similarity to a person who specializes in robbing
banks because that is where the money is. Any hoard invites
exploiters to circumvent the security system. Equally possible,
according to some researchers, is that host and guest evolved
together, one as a giver in moderation, the other as a taker
with a high degree of self-restraint. Probably both routes have
been followed by different insects in the past.

Ants themselves are opportunists, eager to benefit from any
encounter. Many kinds search out aphids and defend them
while getting droplets of honeydew. Some ants carry the
aphids underground and place them on the roots of appropriate plants. There the sucking insects are well protected, while
the ants miss no drop of honeydew day or night. A comparable
behavior of ants in England and adjacent Europe seems more
remarkable because the producer of the honeydew is the caterpillar of the rare blue butterfly. The ants carefully transport
the caterpillars underground, tend them until each is about to
pupate and transform to winged splendor, then give the caterpillar access to the outside world in which the butterfly can
mate and reproduce its kind.

Biting ants and termites, like insects that bear effective stingers as adults, may be good neighbors to animals that are immune or ignored for some reason. A surprising array of different reptiles and birds associate voluntarily with these social
insects, and with others that are similarly armed, apparently as
tolerated guests. We know of almost no other site in which to
find tropical worm lizards (members of family Amphisbae-
nidae), than the underground nests of ants and termites, which
they use as pantries and incubators.

The Nile monitor lizard generally relies upon insects to
provide security for her developing offspring. This two- to
three-foot reptile often takes chances to raid crocodile nests
for eggs and newly hatched young. The monitor herself waits
until rainy season, when the hard earthen walls of termite
mounds have softened by soaking up moisture. Then the vigorous reptile can excavate a tunnel straight in, enlarge the
inner blind end, and have a nest cavity. Soldier termites do
their best to nip at the monitor's scaly skin. Members of the
nasute caste spray the intruder with chemical repellent, to no
avail. The monitor lays her eggs and departs. Termites repair
their nest without disturbing the eggs, within which young
monitors are already developing. Space left around the eggs
connects to termite galleries, admitting oxygen and maintaining the relative humidity at a level almost perfect for the monitor embryos. They will not be ready to hatch until a year later,
when rain again softens the termite mound and facilitates their
escape. Out come the monitors to face predators for the first
time. Often another pregnant female lizard arrives before the
termites can repair their mound, and installs a new batch of
reptile guests where the insects will protect them.

In tropical America, birds benefit from proximity to formidable wasps, each fully two inches in length. The nests are
separated, but by only a short distance. The birds are crow-
sized relatives of the orioles we see building pendant nests of
plant fibers and string in temperate regions; in the tropics they
are known as oropendolas and caciques. Their nests are gigan-
tic but of the same design, hanging as much as six feet long
below sturdy branches of the nearest tree. Somehow the birds
manage to come and go without exciting the insects. The
wasps gain nothing from having the birds so close, yet rush

forth with stingers ready if a toucan with a tremendous beak or some other creature tries to raid the pendant nest for eggs or nestlings. Strangely, the birds seem unable to defend their nests. They and their young lose out if, for some reason, their protective wasps abandon the nursery neighborhood.

SOCIAL INSECTS, SOCIAL PEOPLE

All of us live closer to insects than we usually notice. Our lives and theirs are bound up together. Humankind is the relative newcomer in a world long dominated and subdivided by insects, and in a real sense, we are their guests. We partake of the seeds and fruits that plants elaborate with the unwitting aid of insect pollinators. We show no reluctance to rob honeybees of their delectable honey, or to kill countless silkworms as soon as they have spun their cocoons of shiny strands that human fingers can wind up and use. We exploit the six-legged creatures in every way that we can devise, giving next to nothing in return unless, like ants with their aphid "cows," we offer some physical protection to the beehive or to the mulberry bush upon which the silkworms feed.

In most instances where human activities have contributed to the benefits of insects, their gain has come from receiving free transportation unintentionally from one continent to another, past an ocean barrier the insect was unlikely to cross on its own. Thrown out of ecological context, the displaced species sometimes becomes a pest. We try to correct our error by spreading poisons ruthlessly, and kill more friends, of whose existence we were barely aware, than the pests that are our target. Rapidly the pest adjusts to its new location and the poisons it encounters; soon it is almost immune to substances the chemist has devised and to many not yet formulated. With large families and, often, several generations in a year, an insect is superbly ready to evolve rapidly and exploit an opportunity its ancestors never met.

We introduce exotic crop plants to native insects, and feel dismay when the leaf beetle that maintained a modest population, with grubs feeding on wild vegetation of the nightshade family, discovers more luxuriant foliage on potato plants. It becomes the Colorado potato beetle, and spreads back along the fields of preferred resource until it gets to Europe. Or we manage a northern forest in the Northeast or Northwest to yield the largest possible crop of a chosen kind of tree, and then resent the mother budworm moths that find this unnatural bonanza. We compare the cash value of a theoretical forest unchewed by these insects, the losses they are causing, and the cost of having pilots spray the trees with insecticide from aircraft. We forget that the forest has never been insect-free and never will be, that it has many other values to wildlife and to people. Our profit-and-loss forecast does not include the contamination of pure water seeping from the forest into drainage streams, or the effect of the poison on aquatic insects, and then on the fish that eat them. Who, in our society, has the most powerful lobby where decisions are made—the forest industry, the fisheries people and sports fishers, the advocates of outdoor diversity, or the Environmental Protection Agency?

In promoting diversity, the insects are on our side. Or, better, we might well join them. Over an enormous span of time they have participated actively in developing new opportunities. The human species during far fewer years has worked to simplify and eliminate, as though ultimately to control. We have turned forests into open fields, lush grasslands into arid plains, short grasslands into deserts. Not even the land plants and the land insects, which together colonized the world beyond the wet borders of swamp and marsh, can prevent or patch these rapid changes. The self-centered human guests show little of the moderation that a rove beetle maintains after it has moved into the nest of ant or termite. "To the victor belong the spoils," we say, without looking ahead to wonder what the victors will do when the only world they know is spoiled.

We can scarcely modify the human enterprise to match the methods and follow the behavior of social honeybees weighing forty-five hundred to the pound. Their size and their mode of inheritance constrain them in one direction, ours in other ways. Our behavior, unlike theirs, can be planned, intentional, intelligent. We can examine the actions of unsocial and social insects, of animals of many other kinds, and choose a composite as the model for our own. Our plan can aim for perpetuation of the insect world and ours, a goal the insects unwittingly share with the most responsible in our society.

If we need a motto as a reminder of our purpose, it might well be "The greatest good for the greatest diversity of life," not for the greatest number of any single kind. The insects demonstrate in their behavior how this can be accomplished. Our social groups could look more realistically for a long future by promoting the stability that diversity confers.

Index

Page numbers in *italics* indicate illustrations

FRIENDS IN THE PARK

BY ROCHELLE BUNNETT
PHOTOGRAPHS BY CARL SAHLHOFF

FRIENDS IN THE PARK

BY ROCHELLE BUNNETT
PHOTOGRAPHS BY CARL SAHLHOFF

CHECKERBOARD PRESS
NEW YORK

Published by Checkerboard Press, Inc.
30 Vesey Street, New York, NY 10007.

ISBN: 1-56288-347-X
Library of Congress Catalog Card Number: 92-73844
Printed in Singapore
10 9 8 7 6 5 4 3 2 1   (F 15/15)

This book is dedicated to Anne Brown,
who has touched so many young lives.

Come with me to the park and meet some of my friends.

Here we go for a ride down the slide.
One, two, three. Whee!

We like to play all sorts of games.

When I'm in my wheelchair,
I like to bat the ball with my
racket—"whoosh!"

Throw the ball ...
and make a basket.

My walker helps me stand up and get around.
Watch me kick the ball as far as I can.

Up on my tiptoes reaching up high....

One more time!

Swings are also lots of fun.

My friend needs my help, so I give him a gentle push.

I swing back and forth, going HIGHER and HIGHER and HIGHER.

Bubbles, bubbles everywhere.

POP!

POP!

POP!

18

Whirling and twirling our wonderful wands.

What colors of the rainbow do you see?

Big trucks,

little trucks,

racing trikes,
and motorbikes.

Hold on tight. Here we go up and down, up and down.

Now we go around and around
on the merry-go-round.

Through the tunnel, we wiggle and squiggle our bodies.

It's juice time. Everyone is as thirsty as can be.

Would you
please pour
another
cup of juice
for me?

Ready or not, HERE I COME!

Take my hand.
Let's join our
friends in a
game of follow
the leader.

Your turn.

My turn.

32

Thanks for making our day in the park so much fun!

A heartfelt thanks to all the children and their families who participated in the *Friends in the Park* project. Many of these children have received special intervention services from the Whatcom Center for Early Learning, located in Bellingham, Washington.

The children have told us a bit about themselves:

Anna likes looking at books, going to the zoo, and running in the park. (She appears on page 18.)

Annie likes to scribble with fat crayons, eat applesauce, and chase her dog, Winston. (7, 12, 22, 23, 25, 26, and 27)

Brianna likes storytime, playing dress-up with big hats, and taking care of her baby dolls. (15, 19, 23, 30, 31, 32, and 33)

Brooke likes putting puzzles together, playing dress-up, and giving others hugs and kisses. Brooke has Down syndrome. (11, 13, 18, 22, 25, and 28)

Corbin likes going for walks, making scrambled eggs, and being with his older sister. Corbin was born with spina bifida, club feet, and hydrocephalus. (Cover, 7, 10, 11, 12, 13, 18, 25)

Eddie likes to play with pots and pans, open and close drawers, and walk to the barn with his dad. Eddie has Down syndrome. (16, 23, and 26)

Elliot likes to wash dishes while standing on a big pot, make up songs, and watch for big trucks and tractors. (15, 21, 29, and 33)

Erin likes talking on the telephone, running as far and fast as she can, and playing with the big kids in the neighborhood. Erin has cerebral palsy. (7, 16, 20, 23, 26, and 27)

Felesha likes turning water on and off, playing the piano, and pushing the keys on her daddy's computer. (15 and 32)

Joseph N. likes eating potato chips, riding in his grandpa's truck, and playing in the bathtub. Joseph has a rare metabolic disorder. (28)

Joseph S. likes Winnie-the-Pooh, eating peanut butter and jelly sandwiches, and playing with cars and trucks. Joseph has Apert's syndrome. (20, 24, and 31)

Karen likes to chase butterflies, walk her dog named Phoebe, and climb on everything. (15, 18, 19, 21, 23, 29, 30, and 31)

Kevin F. likes to ride his bike down hills, play with his train set, and climb as high as he can. (15, 19, and 31)

Kevin R. likes to wrestle with his older brother, look at books, and moo like the cows. Kevin has Down syndrome. (15, 18, 19, 21, 23, and 33)

Kyle R. likes listening to his tape recorder, playing baseball in his backyard, and playing in his sandbox. Kyle has Larsen's syndrome. (Cover, 7, 12, 13, and 25)

Kyle S. likes playing with toy dinosaurs, listening to made-up stories, and going to the beach. Kyle was born prematurely. (7 and 20)

Matthew likes playing the piano, blowing bubbles, and eating ice cream. Matthew has Down syndrome. (22)

Meghla likes eating French fries, climbing up and over everything, and cooking. Meghla has cerebral palsy and a seizure disorder. (14 and 17)

Shelby likes riding on tractors, playing peekaboo, and talking like the animals do. (23 and 26)

Walter likes rolling car windows up and down, eating chocolate pudding, and helping his mom clean house. (12, 20, and 28)

*

Rochelle Bunnett received her master's degree in early childhood education from the University of Oregon. For the past twenty years she has been a teacher of young children and a parent educator, and in the last eight years has found a special interest in designing environments and curriculum for infants and toddlers with different abilities. Rochelle makes her home in Bellingham, Washington, with her husband, Rob, and daughter, Annie.

Carl Sahlhoff has been a professional photographer for fifteen years. His work has been published in numerous magazines. Carl lives in Bellingham, Washington, with his wife, Kathy, and children, Joseph and Kira.